CUSTOMER TRANSFORMATION

Customer First!
Data Second!

CUSTOMER TRANSFORMATION

A 7-Stage Strategy for
Customer Alignment and Business Value

CHRIS **HOOD**

CH PUBLISHING
a division of CH Digital Ventures,
Rancho Santa Margarita, California

Copyright © 2023 by Chris Hood. All rights reserved.

Customer Transformation is a registered trademark by Chris Hood.

Printed in the United States of America. No part of this book my be used or reproduced in any manner whatsoever without written permission except in the case of brief quotations embodied in critical articles and reivews, or where otherwise is referenced through original source materials. For permission requests, write to the publisher, addressed "Attention: Permissions Coordinator," at the address below:

CH Publishing
30211 Avenida de las Banderas #200
Rancho Santa Margarita, CA 92688
chpublishing.com

Although every effort has been made to ensure the accuracy and reliability of the information presented in this book, neither the author nor the publisher are liable for any loss or damages, incidental or consequential, that may arise directly or indirectly from the use or application of this information. All names, characters, businesses, places, events, and incidents have been anonymized unless explicitly stated otherwise.

The strategies and advice contained in this book do not guarantee success and the results may vary from individual to individual. The reader is responsible for understanding and accepting the inherent risk that results may differ.

Furthermore, the publisher and author disclaim all liability to the maximum extent permitted by law, in the event that the information, commentary, analysis, opinions, advice, and/or recommendations in this book prove to be inaccurate, incomplete, or unreliable, or result in any investment or other losses. The reader is advised to exercise their judgment and verify information before making any decisions based on the content of this book.

Library of Congress Cataloging-in-Publication Data

Names: Hood, Chris, author.
Title: Customer Transformation: A 7-Stage Strategy for Customer Alignment and Business Value.
Description: First Edition. | California : CH Publishing, 2023.
Identifiers: LCCN 2023911389
  ISBN 979-8-9885384-0-0 (HC) | 979-8-9885384-1-7 (PB)
Subjects: LCSH: Customer Relations | Strategic Planning | Leadership (Business)
LC record available at https://lccn.loc.gov/2023911389

To my dad,
the inspiration for all of my best ideas.

TABLE OF CONTENTS

Introduction   11

STAGE ONE: CUSTOMER
    1. Foundations & Aspirations   21
    2. The Power of Ideation   34
    Action Plan One   50

STAGE TWO: INTERFACES
    3. People Interfaces   55
    4. Artificial Customer Relationships   67
    Action Plan Two   85

STAGE THREE: JOURNEYS
    5. The Outside-In Perspective   91
    6. Experiences in the Moment   105
    Action Plan Three   119

STAGE FOUR: ECOSYSTEM
    7. Channel-less Innovation   125
    8. Communities at Scale   141
    Action Plan Four   161

STAGE FIVE: CULTURE
    9. Cultures of Praise   167
    10. Delivering Inspiration   184
    Action Plan Five   195

STAGE SIX: TECHNOLOGY
    11. Purpose-Driven Technology   201
    12. Data-Driven Responsibility   217
    Action Plan Six   230

STAGE SEVEN: BUSINESS
    13. The Value of Obsessed Leadership   235
    14. Customer Value Alignment   253
    Action Plan Seven   274

Endnotes   279
Index   296

Customer Transformation - A business strategy through which a company *transforms* its processes, culture, and technologies to *align* with its customers' *ever-evolving* needs and aspirations.

ZERO

INTRODUCTION

Moments of change occur around us each day, and in that spirit this book begins with a moment of closure and reflection—a day of transformation, if you will—at a place I loved and once called my second home: Cinemapolis, a movie theater nestled in the heart of Anaheim Hills, California, where I obtained my first full-time job more than 30 years ago.

Imagine stepping into a world filled with larger-than-life tales, popcorn aroma wafting through the air, hushed anticipation, and the resonating echo of the simple yet enthralling words, "...coming soon." Cinemapolis offered this experience for years—a world away from the real world, a shared dream that brought the community together as patrons embarked on cinematic adventures. As of this week, while I'm writing this, the theater closed its doors, and the landlord will usher in a Tesla showroom to fill the space. I can't help but contemplate whether this signifies the loss of a historical structure in my life or the dawn of a new era.

What began as selling candy and cleaning auditoriums

at age 17 ultimately taught me the essential values I still heed today. While working at the theater, I first learned a well-known theory: "The customer is always right." Thirty-five years later, the depth of this statement is more nuanced.

The common business mantra is meant to instill a customer-centric culture, promoting a service ethos that prioritizes customer needs, satisfaction, and loyalty. It's about treating customers with respect, listening to their concerns, and making genuine efforts to solve their problems.

However, in reality, the customer is not always correct or reasonable. Customers sometimes have unrealistic expectations, make mistakes, or exhibit unfair demands. On the flip side, in recent years, I've witnessed leaders who believe they know better than their customers or have imposed their personal biases on them, resulting in a substantial negative impact on their company's value.

But while one customer might not always be right, the collective voice of customers can provide invaluable feedback. If many customers express similar complaints or suggestions, it's indicative that some aspect of the business might need to be revised or improved. This alignment of your core customer voice is, and always will be, right.

Another guiding principle at the theater was "anticipate the guests' needs" before asked. These philosophies, now planted in customer call centers and hospitality companies worldwide, were not just words; they were a compass, directing every interaction and decision we made at the cinema. But how exactly do you predict customers' needs? Data? Personalization? Those annoying web cookies? Fortune 500 companies have been asking this question for the last two decades, and even with the onslaught of artificial intelligence, the belief is that we're closer than ever to reaching it.

The reality is that change is constant. Customers' needs, expectations, devices, engagement, opinions, and beliefs are constantly in flux next to a business' products, services, locations, prices, and reputation. Understanding this, and being able to adapt and serve customers in the moment, aligned with core values, is the heart of customer transformation.

These principles of customer focus were more than mere words on a training sheet for me to agree to and forget all those years ago. Instead, they seeped into my core, becoming second nature to me—like a reflex, instinctive and natural. This philosophy became a passion I carried into every job after Cinemapolis. On the surface, this may feel like a given. However, I challenge you to set aside your assumptions and position your mindset with the enduring lesson: without customers, your business doesn't exist.

I forged many memories in the theater's comfortable darkness—first dates, hosting classic film series, my children's first movies, enjoying popcorn with Gwen Stefani, interviewing Kathy Bates, and sneaking Sean Connery into one of his films. It was more than just a workplace; it was a hub for cultivating lifelong friendships, honing skills, and shaping experiences. For 15 years, I shared in the joys of co-workers' graduations, marriage proposals, and the birth of their children, interviewed and trained hundreds of new employees for their first jobs, and tackled everyday customer challenges that became woven into the tapestry of my life and became a narrative of change and transformation.

As I reflect on the arc of my professional journey, it is increasingly clear that the philosophy of customer alignment has been a steadfast pillar. From my formative years at Cinemapolis to every professional endeavor that followed, the commitment to prioritize the customer remains.

This book covers seven key stages that serve as the cornerstone for fostering thriving customer relationships and enhancing business value. Each stage of the Customer Transformation framework outlines how to shift your organization's mindset toward a customer-centric focus, accompanied by engaging stories, business case studies, and practical examples. The seven stages are:

1. **Customer:** focuses on understanding customers' needs, expectations, and evolving behaviors
2. **Interfaces:** explores the touch points and connections between customers and your business and how innovation creates new people interfaces
3. **Journeys:** investigates the customer experience from start to finish, highlighting how customers engage moment-by-moment in a digital world
4. **Community:** discusses building and nurturing a community of engaged and loyal customers at scale
5. **Culture:** illuminates the significance of fostering a customer-centric culture within your organization and how it can impact how your business functions and interacts with its customer base
6. **Technology:** looks at the role of technology and data in meeting and anticipating customer needs and enhancing a company's purpose
7. **Value:** focuses on leadership and customer alignment to quickly adapt to changing markets and significantly boost business value.

See the framework's value chain on the next page.

CUSTOMER TRANSFORMATION

FRAMEWORK

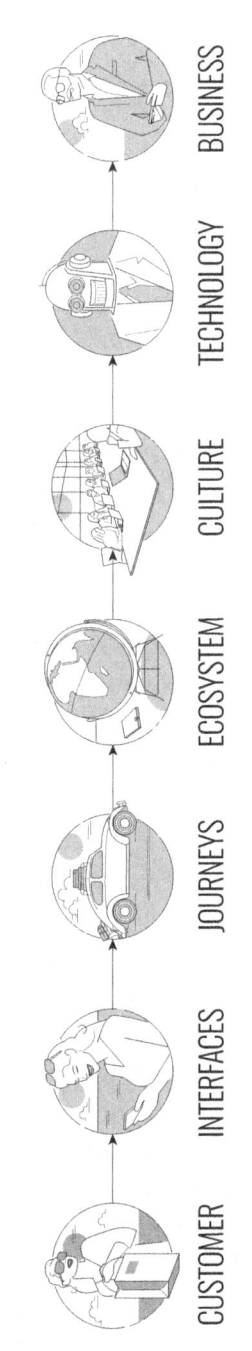

CUSTOMER → INTERFACES → JOURNEYS → ECOSYSTEM → CULTURE → TECHNOLOGY → BUSINESS

OUTSIDE YOUR COMPANY ---------- INSIDE YOUR COMPANY

Action Plans

Following each stage, you will find an action plan with next steps and goals to help implement the insights from the book. To assist in the process, here is how each action plan is structured:

> **Introduction:** The introduction provides an overview of the stage you're about to begin. It presents the key concepts, topics, and the overall objective of the stage. Read this carefully to understand the upcoming activities' context and purpose.
> **Day 0: Reflection and Goal** - A few days before you start the activities, take some time to reflect on the given statement. This reflection should be your starting point, shaping your initial perspective. The goal introduces what you should aspire to achieve in this stage. This sets the mindset you should adopt and the objectives to strive for during the stage.
> **Day 1: Workshop** - This is your first team meeting. You'll be introduced to this stage's materials, processes, and themes. Your team will begin exploring the concepts presented and connect them to your company's unique circumstances.
> **Day 1: Workshop Questions** - These questions stimulate discussion and thought during your first team meeting. They serve as a guide to navigating your exploration of the introduced concepts. You should aim to answer these questions based on your company's specific situation and objectives.
> **Day 2 - 7: Homework and Next Steps** - After your first meeting, the team should reflect on the initial discussion and start working on the tasks identified. This could involve analyzing your current situation,

researching solutions, or outlining strategies. Maintaining open communication within your team during this period is crucial to share insights, updates, and challenges.

Day 8: Touchpoint and Next Steps - You have your second team meeting one week later. You review your progress, discuss the homework, and plan your next steps here. This meeting is about converting your insights into actionable strategies and steps.

Day 9 and Beyond: Action Plan - This section outlines your deliverables for a 30 - 60 - 90-day plan. It is the roadmap for the practical application of your new strategies. It includes the tasks, goals, and metrics to track as you implement changes over the following weeks and months. Each period should build on the last, aiming to drive continuous improvement and adaptation based on results and feedback.

Customer Transformation is about keeping pace with your customers' ever-changing needs and expectations. It's about learning from the past and, more importantly, highlighting how these principles guide us for future growth and business value. As you delve deeper into this journey, I hope my stories will inspire you to embrace the all-important role of the customer in driving your transformation.

STAGE ONE: CUSTOMER

Imagine your Customers' Aspirations

ONE

FOUNDATIONS & ASPIRATIONS

"Think of digital transformation less as a technology project to be finished than as a state of perpetual agility, always ready to evolve for whatever customers want next."

Amit Zavery, VP & Head of Platform, Google Cloud

The Mystery of the Abandoned Carts

A few years ago, the CEO of a clothing e-commerce company hired me to improve their website and develop a digital strategy for business growth. During my first on-site visit, she said, "Chris, we have a problem. We're losing a ton of business because people are abandoning their carts. I've taken a close look at the checkout process, and it's too confusing. I want to simplify the entire process by removing the current payment system, changing the checkout steps, and adding PayPal."

I asked, "Why do you believe the checkout process is causing people to abandon their carts?"

She responded, "Well, it's confusing for me, and I've had others look at it, and they agreed."

I validated her concern and then followed up with another question: "How do we really know that's the main reason for abandoned carts? Do we have any concrete data to back up this idea?"

I proposed incorporating a simple survey and a coupon for the customer's time. The survey would pop up or be sent via email when they left their cart, and every respondent would receive a 10%-off coupon to coax them back to the site and complete their purchase.

With a grimace, the CEO responded, "I don't want to post a survey; they're cheesy. Please just change the checkout process, and we'll see if that helps."

Even though this conversation was awkward for both of us, I pushed back one final time and explained, "I promise you no one will think a survey is cheesy, and it will give you valuable information about your customers."

She hesitated, but then finally agreed for me to set up the survey and coupon I had proposed. It offered five possible explanations, from the consumer's point of view, as to why they had abandoned their cart before completing the purchase:

1. The checkout process was too confusing.
2. I couldn't find enough information about the product.
3. The price of the product was too high.
4. You didn't have the size I was looking for.
5. I changed my mind, or for another reason.

We collected more than 1,000 responses, and the results were clear—but not what the CEO had expected. Nearly 78% of these consumers indicated that *the product was too expensive*. When I presented these findings to the

CEO, she refused to accept them. "We're not changing the price of our products," she insisted. "Fix the checkout process."

I reluctantly backed down and agreed to her demand, but it still bothered me. Despite clear data, she had let her bias cloud her judgment. It didn't matter that the consumers had spoken; *she* knew better.

Unfortunately, within eight months this CEO's company had gone offline.

There may have been other factors that played a part in the eventual failure of the company, but one thing is certain: This CEO might have been able to avert her company's demise but missed the opportunity because her focus was on the technology, or possibly her pride, instead of the customer.

I've frequently seen this same mistake. In this heady environment of ongoing technological evolution, we hear a lot about *digital transformation*, often at the expense of another, perhaps even more important, a concept that I call *customer transformation*. These concepts are closely connected (and the latter is the main subject of this book), so let's take some time here to understand their similarities and differences.

Digital vs. Customer Transformation

Let's start with digital transformation. Here's Gregory Vial's, (Department of Information Technology, HEC Montreal) definition:

> *a process that aims to improve an entity by triggering significant changes to its properties through combinations of information, computing, communication, and connectivity technologies.*

This definition is very academic—which isn't a bad thing—but the focus is almost exclusively on the data and technology, while the business itself gets only a slight nod as "an entity," and the customers aren't mentioned at all. As a result, it's not really clear why or to what end we need "to improve" the "entity" in question. Are we embracing technology for technology's sake or out of a vague fear of being left behind?

To be fair, I don't know any CEOs, CTOs, or CIOs who would be comfortable saying they're "embracing technology just for technology's sake," but it's still surprising how many individuals in executive positions are unable to clearly articulate why their company needs a digital transformation. What value, exactly, is the digitization process expected to bring?

Even when executives tout specific values of a digital transformation, they tend to view them through a rather narrow company lens, in terms of improved efficiency and value-generation that digitalization brings to *the company* (the ways digitalization affects the company's processes, products, and services). While this kind of focus on value is certainly an improvement over the vague, valueless perspective featured above, it remains a partial truth if we don't directly tie the value of digitalization back to the one major participant: the customer, who everyone is aware of but often goes unmentioned.

With that in mind, consider this somewhat more well-rounded definition of digital transformation by Bill Schmarzo, CTO of Dell EMC Services:

> *Digital Transformation is the application of digital capabilities to processes, products, and assets to improve efficiency, enhance customer value, manage risk, and*

uncover new monetization opportunities.

Schmarzo's definition stands out for its emphasis on the ability of digitalization to improve value, including the value that customers hold toward a company's offerings. In that regard, Schmarzo's definition represents an important step in the right direction, a step that has been taken by other observers as well (e.g., McDonald and Wren).

But I am convinced that we need to take this a step further. When technology and business leaders first discussed the concept of digital transformation in company break rooms, email threads, and strategy sessions, the focus was squarely on *the customer*. Saul Berman, VP and Partner of IBM Global Business Services, nicely captured that original perspective was back in 2012:

> [C]ompanies seeking opportunities in an era of constant customer connectivity focus on two complementary activities: reshaping customer value propositions and transforming their operations using digital technologies for greater customer interaction and collaboration . . . [B]usinesses aiming to generate new customer value propositions or transform their operating models need to develop a new portfolio of capabilities for flexibility and responsiveness to fast-changing customer requirements . . . [E]ngaging with customers at every point where value is created is what differentiates a customer-centered business from one that simply targets customers well.

Implicit in Berman's remarks is the recognition that it was the customers who—due to their rapidly-increasing exposure to new technologies and the insatiable need for this exposure—spurred faster, more useful, and more satisfying technologies. They were driving industries-wide shifts to digitalization. Because those customer needs were quickly and constantly evolving, Berman and others recognized that the nature of digitalization and our understanding of the value it brings would also need to continue evolving.

It is this customer focus that has been largely lost in our current discussions of digital transformation. We can credit this loss to more than 7 years of countless energetic pitches for digital services and cloud technologies by savvy sales and marketing teams. These pitches have included tantalizing statements such as the following (all from actual marketing messages):

- "Embrace the future of business with cloud-powered digital transformation."
- "We can help you unlock the full potential of your business with data-driven digital transformation."
- "Digital transformation happens with an agile content management system built for speed and scale."
- "Your organization can use IoT tools to accelerate digital transformation."
- "Invest in artificial intelligence, and see your business reach digital transformation success."

The problem is none of these sales statements make any mention at all of the customer. This brings us to the concept of *customer transformation*. In several articles I wrote in 2016, ("A Customer Transformation Mindset," "The Ultimate Customer Transformation Strategy," "Four Secrets About Customer Transformation"), I coined this term to restore needed focus on the evolution of a company's interactions with its customers for the purpose of better meeting their needs and aspirations. The term also conveys recognition of the fact that customer needs and expectations are not static but rather constantly transforming along with the ever-changing marketplace. To a truly responsive company, this ongoing process of customer transformation can call for changes to the company's products, services, customer engagement strategies, and even organizational structure. The objective of customer transformation is to enhance customer satisfaction, foster loyalty, and increase lifetime value by creating a more efficient and fulfilling customer journey.

Customer transformation, especially within the current environment of extensive and aggressive digitalization, largely overlaps or can be considered a subset of digital transformation, which involves integrating technology into all areas of your business in ways that support both processes and people. Effective digitalization, then, will alter the value you provide to your staff, partners, and customers. A focus on customer transformation can be seen as a safeguard against losing your way amid technological advancement: Customer transformation reorients your team's efforts back to the human element and keeps them focused on what is arguably your business's central source of value: your customers.

Another way to understand the difference between common instances of digital transformation and customer transformation is to look at starting points: Digital transformation often begins when a company thinks about desired technology changes. Customer transformation, in contrast, starts by considering customer aspirations regarding technological engagement. The distinction is in the terms themselves; digital transformation starts with "digital," and customer transformation starts with "customer." The transformation comes next.

This is the difference between inside-out and outside-in perspectives: The inside-out perspective so common today with digital transformations gives primacy to the digitization occurring within the company and gives only secondary consideration to the customers' "outside" perspective regarding the transformation's value to the business. For this reason, digitalization is frequently seen by executives more as a cost to be borne than as a source of ongoing business reward. Customer transformation instead takes an outside-in view, empathizing with the "outsiders" (i.e., customers) and giving primacy to their perspectives. I'll discuss these concepts in our next chapter.

It may be useful to clarify that the term "customer transformation" isn't meant to imply that you are transforming your customers (although it's certainly true that your marketing, and quality of customer service, will affect them). Instead, the term stresses the importance of recognizing that your customers' expectations and needs are constantly changing due to a multitude of factors in a constantly evolving digital marketplace. Your ability to provide value to your customers on an ongoing basis will depend directly on your ability to evolve your company's technologies, products, services, and experiences with those customers in mind. The term "customer transformation," therefore, actually suggests parallel transformations: a recognition of the continually transforming customer and empathy with whom motivates and informs a continual transformation of your business.

The Whys and Wherefores

We now have the concepts in play to better understand how the CEO, I attempted to help, misunderstood the abandoned shopping carts situation. She was so focused on the technology that she allowed the transformation of her business to be guided by her own expectations of digital transformation rather than by an objective assessment of where her customers were in their transformative journey. The CEO assumed the problem to be inherently technological, when it was actually a misunderstanding of her products' exchange value, and the tragedy is that she was provided with customer data that could have corrected this, but she chose to ignore it or worse, not believe it. Because she lacked a customer transformation mindset, she neither sought out the relevant data nor gave that data serious consideration once she had it.

Data is crucial to customer transformation. It is your ability to leverage data documenting your customers' changing expectations, needs, and aspirations that will determine your ability to keep the transformation of your business in sync with the transformation of your customers. Good customer data is how you answer the all-important question at the heart of both digital and customer transformation: Why does your company need this new technology?

I've sat in dozens of board rooms or workshops with C-level executives who've stated adamantly that they are going to bring in X technology, or develop Y service, or go to Z cloud. But when I ask *why* they've chosen this particular digital solution and how it maps to the needs of their customers, most of these executives either recite an internally focused mandate, a personally biased belief or goal, or they simply stutter and stammer their way through a non-answer.

I once stated to a room full of overly-confident executives, "If you believe you know what your customers want before asking them, you're most likely going to lose those customers." Few of us are as clever as we think we are, and when it comes to our customers, it's easy to proceed with false assumptions. Just like any significant relationship: It doesn't stand much of a chance if you don't actively *listen*.

The other reason you're likely to lose customers regardless of which sophisticated new technology you bring in, is the fact that customers are constantly changing. I suspect this is truer now than it's ever been in free-market history, simply because the pace of contemporary technological change is so breathtaking.

This is the foundation of customer transformation: *You must be able to connect every technology decision you make, back to one or more specific, documentable customer value propositions. Always.*

Think of it this way: A Boeing 747 is made up of roughly 6 million parts. Imagine sitting down with a group of passengers and explaining that their impending Boeing 747 flight was missing one important screw, but we were going to proceed with the flight anyway (perhaps with the offer of free drinks down through economy class as compensation). Do you think some of the passengers would insist on changing their flight? Of course they would. Why? *Because every part of the airplane (known or unknown) has a purpose and perceived value to the passenger.*

Your business—including every part or process in it that has any direct or indirect effect on the experiences of your customers—is the airplane. Your customers are the passengers. The more mismatches there are between your business structure, practices, and services, and the needs and aspirations of your customers, the more likely your customers will be in for a bumpy ride—and the more likely they'll bail on you (just like with those abandoned shopping carts).

I participated in a group conversation on LinkedIn about productizing APIs (Application Programming Interfaces). One individual claimed that a business could profitably monetize the APIs to generate new business opportunities. Another participant disagreed and said, "But if you productize your APIs, you'll be disregarding integrations, which account for 98% of what APIs are used for." My response challenged both of their positions: "Yes, but if you can't explain how 100% of those APIs are being used to drive customer satisfaction, then you don't need the APIs in the first place."

Most organizations seem to have this unspoken attitude: *we simply need the technology to be there*. Very few can explain how every element, down to the interfaces of the technology stack, can grow business value and customer satisfaction. But when you begin by empathizing with your customers, really listening to them so you can understand them, you can develop strategies, embrace innovations, and adopt carefully targeted technologies that have genuine value impact for your customers.

Blockbuster's Big Bust

It's a cold, hard truth of business success: *If you fail to continuously adapt along with your customers, you will (eventually) go out of business.* This truth holds 100% of the time, regardless of how big your company might be or what past successes it has achieved.

One of the most well-known examples of a large company that went under because it failed to adapt is Blockbuster, which in the late-'90s owned more than 9,000 video-rental stores and boasted 65 million registered customers. Yet, by 2010 Blockbuster had filed for bankruptcy, with its last corporate store shuttering just four years later. So what happened?

Most people claim that Blockbuster failed because of technology disruption, while others point to executive hubris and greed. While there is an important element of truth in both these claims, the reality tying them together is that Blockbuster went bust because it didn't listen to its 65-million-plus customers. Yes, technology was disrupting the market, but more importantly, these technological advancements were transforming Blockbuster's customers, who were growing increasingly hungry for the convenience and new possibilities being offered by the online world. Yet, as Ross notes, "[t]he company seemed unwilling to even contemplate the fast-evolving shift in consumer behavior." And, by all accounts, Blockbuster's executive lineup included one or more arrogant individuals who had succeeded in making tons of money from the in-store experience and weren't about to let go of their late-fee profits. These profits totaled $800 million in one year alone, nearly as much, ironically, as the $900 million in debt the company carried at the time of its bankruptcy filing. But the real significance of this overconfidence and greed is that it made Blockbuster's leadership deaf to its customers' complaints and increasing frustration with the company's late fees. As a result, Blockbuster continued to offer their "Come-to-Us' business model instead of "We'll Come to You," and this insensitivity to customer transformation led to the company's demise.

The reality is, once an organization has fully digitized, its customers will still evolve and transform the ways they engage with that organization. Customer transformation and digital transformation are, thus, both never-ending processes. Arguably, this is the most important reason why the majority of organizations—more than 70%—fail at digital transformation. Too many decision makers view digitalization, like many other aspects of corporate development, as more or less a once-and-done phenomenon. In this mindset, building the company's cloud platform, for example, is treated the same basic way as erecting a new corporate office. All construction is finished in time for the ribbon-cutting ceremony, not to be repeated (apart from minor repairs and perhaps eventual renovations) for as many years as the new building can be made to productively last. This way of seeing things betrays a blindness to the company's customers, who will continue to transform from month to month and year to year. Sensitivity to a customer's transformation mindset requires the "state of perpetual agility" advised by Amit Zavery in the quotation at the beginning of this chapter.

TWO

THE POWER OF IDEATION

*"The world as we have created it is a process of our thinking.
It cannot be changed without changing our thinking."*

Albert Einstein

The Winnebago Rental
"It sure would be nice if we could rent a Winnebago and do our assessments right there in the middle of their neighborhood."

While meant as a joke, this Golden Gate Regional Center (GGRC) brainstorming suggestion, made while participating in a 2013 session with other GGRC staffers and with students from Stanford University's Hasso Plattner Institute of Design, would soon grow legs. As this Harvard Business Review article chronicles, GGRC had been partnering with the Institute to come up with creative ways to streamline and improve GGRC's complicated assessment process for people with developmental disabilities in the San Francisco Bay Area seeking services and financial support. However, "the system was not designed to make life easy for clients or staff members," and the assessments typically involved three or more months of parents dragging "their children to a daunting series of meetings and examinations in unfamiliar places, which distressed the children and led many parents to abandon the process."

But in the spirit of discovery, the staff member's humorous suggestion from that brainstorming session became a reality, and on February 14, 2014, two Stanford students and eight GGRC staff members drove a rented Winnebago from San Mateo's GGRC office into the neighborhoods of potential clients as an experiment to determine whether such a 'field' approach might be a workable alternative to in-office assessments. The results surprised everyone: nine assessments in less than two hours.

Even though assessment processing took "another week of concentrated work at the GGRC office," the experiment was an undeniable success since "that was 10 weeks faster than usual." The GGRC staff "had learned valuable lessons about how they could improve the process for all clients."

After further review and consideration given to the expense involved, the GGRC chose not to scale up the Winnebago-based approach. However,

> [T]he exercise wasn't really about the Winnebago; it was about pushing beyond what was comfortable, taking a different point of view, trying something new, experimenting, and discovering what to try next. It was part of a larger effort by GGRC staffers to rethink what was possible for their clients and themselves.

That simultaneously challenging and thrilling process of rethinking what is possible—of envisioning genuinely innovative solutions to our customers' needs in light of what we've discovered about their experienced moments and journeys within the broader context of the ecosystems they inhabit—is called *ideation*, the focus of this chapter.

The Art of the Possible

Key to the success of the Winnebago experiment was the staffers' and students' ability to transcend the "they come to us" mentality that had kept the existing assessment process bound largely within the confines of the office setting. This transcendence was both spatial and conceptual—not only did the experiment literally deliver the assessors to the doorsteps of potential clients, but it also gave visible expression to a genuine service commitment on the part of the staff members and to their willingness to make the process as seamless and stress-free as possible for their clients.

The phrase "art of the possible"—often used in the context of innovation, leadership, and change management—refers to this ability to think outside the box (rent a Winnebago!), push boundaries, and boldly explore new ideas to achieve a desired outcome that transcends existing limitations. It also emphasizes the importance of being proactive, adaptable, and creative when approaching problem-solving; it encourages individuals and organizations to take a forward-thinking approach to decision-making, to embrace risk and uncertainty, and to be open to exploring new and untested ideas and approaches.

In our context of customer transformation more specifically, the art of the possible means *proactively identifying and implementing innovative solutions that meet customers' evolving needs and expectations beyond the current scope of solutions that customers are seeking or of which they may even be aware.*

Big-Picture Thinking

For a lot of people, quick thinking while in front of your customers can be a challenge, but as hinted above, it can be an exciting one. It requires a certain predictive capacity that comes with what might be called "big-picture thinking," or the ability to envision the past, present, and future of the complex ecosystems in which your customers live out their lives. A good example of this is Bill Gates and his development of Windows, which combined instinct, market relevance, demand, rumor mill, and other factors. And, boy, did it pay off.

The predictive aspect of this comes in layers, sandwiched between the steps of an iterative validation process. With a thorough understanding of your customers in mind, you make an educated guess or prediction about what is coming. For example, you might predict that your customers will want to use Virtual Reality. That may mostly just be a biased belief on your part, but it feels right, so you bounce the idea off your colleagues and partners, get feedback, and continue researching the topic. Next, you discover articles suggesting that Meta and others are backing away from their investment into the Metaverse. This validation of your idea—in this case potential negative validation— may make your prediction of customers' coming obsession with VR less enthusiastic. Of course, maybe there are other ways to understand Meta's shifting priorities, so you keep exploring, seeking to grasp the big picture.

In some cases, such as the GGRC brainstorming session, a prediction can lead to a testable hypothesis. While not every prediction will lead to a testable hypothesis, and not every hypothesis will pan out when tested (most won't), it's all part of the ideation and discovery process. When we develop and test hypotheses for our customers, we can build upon each failure and each success with new ideas that get us closer to products and services our customers may want.

I'm not suggesting that everyone can become a master of seeing the big picture. Some individuals will look at the data and tend to see only the most obvious and easy things to understand. Others will let their imaginations far outstrip the data and make ridiculous suggestions, like teleporting products directly to customers' homes. Others will be able to see fairly accurately what the future holds. However, all of us can shift our thinking in this direction and do the research into our customers and the broader market to become aware of trends and adapt our ideas in order to predict what customers' future needs might be.

More than that, embracing the art of the possible and big-picture thinking can be a significant component in businesses' efforts to develop a customer-centric culture that drives transformational change. This may involve leveraging new technologies such as AI and machine learning to deliver personalized experiences and to streamline processes. It may also spur the rethinking of business models and organizational structures so that they better align with customers' values and preferences. Overall, the art of the possible and big-picture thinking empowers businesses to anticipate and respond to customers' changing needs—creating a competitive advantage in today's fast-paced, dynamic marketplace.

10x Thinking

The above strategy is closely related to one of the more exciting ideation perspectives to emerge in recent years: 10x thinking. This framework encourages individuals and organizations to think beyond incremental improvements and instead aim for breakthrough innovations with exponential results of 10 times the impact. How this mindset can make such a real-world difference is explained by the head of Google X, Eric Teller, this way:

> [W]hen you're working to make things 10 percent better, you inevitably focus on the existing tools and assumptions, and on building on top of an existing solution that many people have already spent a lot of time thinking about. Such incremental progress is driven by extra effort, extra money, and extra resources. It's tempting to feel [that] improving things this way means we're being good soldiers, with the grit and perseverance to continue where others may have failed -- but most of the time we find ourselves stuck in the same old slog. But when you aim for a 10x gain, you lean instead on bravery and creativity—the kind that, literally and metaphorically, can put a man on the moon.

Such "moonshot thinking" is necessary in a world that is transforming at an exponential rather than linear rate. As entrepreneur and blogger Marc Winn notes, "[l]inear thinking was really useful when the edge of our universe was our local area. But the world is now global and hyper-connected . . . The real risk to an exponentially changing world is evolution, rather than revolution. That's why it's no longer good enough to simply improve what's there."

When it comes to customer transformation, applying 10x thinking can lead to radical changes that completely transform the way a business interacts with its customers. A good example of this is how the design team at Pepsi radically re-envisioned the soda dispensing machine to create Pepsi Spire, a new type of touch screen fountain machine.

PepsiCo's CEO, Indra Nooyi, explains:

> Other companies with dispensing machines have focused on adding a few more buttons and combinations of flavors. Our design guys essentially said that we're talking about a fundamentally different interaction between consumer and machine. We basically have a gigantic iPad on a futuristic machine that talks to you and invites you to interact with it. It tracks what you buy so that in the future, when you swipe your ID, it reminds you of the flavor combinations you tried last time and suggests new ones. It displays beautiful shots of the product, so when you add lime or cranberry, it actually shows those flavors being added—you experience the infusion of the flavor, as opposed to merely hitting a button and out comes the finished product.

It would have been simpler in nearly every respect—from designing to engineering to manufacturing, etc.—to simply "add a few more buttons" to existing machines, but as Nooyi says, "We don't sell products based on the manufacturing we have but on how our target consumers can fall in love with them."

10x Thinking & Customer Transformation

10x thinking is a powerful approach to ideation. Here are some specific ways you can apply it to customer transformation:

Rethink the customer journey. Instead of simply optimizing individual touchpoints along the customer journey, take a step back and reimagine the customer experience from start to finish. What if you could create a completely frictionless customer journey that delights customers at every step?

Challenge assumptions about customer needs. Rather than assuming you know what your customers want, use 10x thinking to explore what they really need. For example, instead of merely increasing the speed of your customer service response times, ask yourself if there's a way to eliminate the need for customer service altogether by designing products that are so intuitive they never require support.

Leverage emerging technologies. Rather than looking to existing technology solutions to improve the customer experience, use 10x thinking to explore emerging technologies and their potential to revolutionize customer interaction. For example, could virtual reality or augmented reality be used to create a completely immersive and engaging customer experience?

Embrace experimentation and risk-taking. Encourage a culture of innovation where employees are empowered to experiment, learn from failures, and iterate their ideas. Be willing to take calculated risks and invest in groundbreaking projects that could potentially deliver exponential returns.

Explore new business models. 10x thinking can also be applied to business-model innovation. What if you could disrupt your industry by creating a new business model that completely upends the status quo? For example, could you create a subscription-based model that eliminates the need for customers to purchase individual products?

A caveat: As with every other approach that promises groundbreaking profits by shaking up the status quo, it is disturbingly easy for some proponents of 10x thinking to lose customer focus while in pursuit of the "wow" factor. One example is from Chuck Gulledge, in which 10x thinking is defined in terms of the potential benefits brought not to the customer but to the entrepreneur, who is asked to envision "10 times your income as an individual or 10 times your revenue as an organization." In fact, nowhere in the article are the needs of customers mentioned, and even the title of the article itself, "What a '10x Mindset' Can Do for You and Your Team," describes a business-centric rather than customer-centric focus.

Empathy and Design Thinking

This style of innovation requires genuine empathy toward your customers, the ability to put yourself in the customer's shoes and see things from their viewpoint. Such empathy is foundational to the final influential ideation perspective we want to focus on in this present chapter: Design Thinking. One of the leading proponents of this approach, Tim Brown, puts it this way: "Empathy is at the heart of design. Without the understanding of what others see, feel, and experience, design is a pointless task."

Design thinking is a human-centered, iterative problem-solving approach that can be applied to customer transformation to create innovative solutions and improve customer experiences. According to Google's Creative Skills for Innovation Lab, design thinking utilizes empathy, expansive thinking (like 10x), and experimentation to find the most effective ways to meet customers' needs and expectations.

A classic example of design thinking is Bank of America's "Keep the Change" program origin story. As Karen von Schmieden recounts, BOA was exploring ways to more effectively reach the "boomer-age women with children" demographic. As a first step, the BOA team (assisted by design and consulting firm IDEO) conducted observations and interviewed families and individuals across the U.S. in order to document their spending and banking habits. This exercise in empathy led to some interesting discoveries, including that mothers were often in charge of household finances.

Many of these women kept checkbook registries, in which they listed their bills, expenses and ATM withdrawals. In the beginning of the 2000s, this was a common way to manage money – a time before banking apps and online banking was widespread. When the design researchers looked at the checkbook of a woman in Atlanta, they realized that she rounded up numbers: Instead of listing [$22.73] for a refueling bill, she listed 23 dollars, for example. This simplified calculations, but it also added a little buffer each month – at the end, a few unexpected dollars were left over for her. A second, important realization concerned the inability of mothers to save. Some of them simply did not have enough money to put anything aside, others could not control their impulse buying. The researchers realized that this behavior revealed an unaddressed need in banking and offered the opportunity to create value for customers and the bank.

With these realizations in mind, BOA assembled a team to brainstorm novel product concepts to address these particular customers' needs. After much discussion, the team settled on a favorite suggestion, the "Keep the Change" program, which would "round up all purchases made with a debit card" and transfer the surplus to a savings account. The team put this idea to the test by creating a cartoon video that illustrated the rounding up service; they then carried out an online survey in which 1,600 individuals reported their reactions to the video. The response was overwhelmingly positive, and a modified version of the program was offered by BOA to its customers in September 2005. Data indicated that "99% of customers who signed up for the program stayed with it."

Commenting on the impact of the Keep the Change service, Faith Tucker, Sr. VP and Product Developer at Bank of America during the time of the project, had this to say:

> There was an almost unexpected and very emotional effect from this new service ... people who previously never had savings suddenly did ... and it wasn't the amount that mattered; even a small amount of money in their savings account gave them a sense of power and control over their finances.

The central role of empathy in design thinking is consistently heralded by the approach's proponents, and stories abound of how radically new ideas were born out of insights gained through time spent in deep exposure to a business's target customers. Eli Woolery of InVision tells the story of a team from Stanford's "d.school" that was tasked with coming up with a design project that would help solve the problem of infant mortality due to hypothermia in developing countries.

> Initially, the design team thought redesigning existing hospital incubators to be simpler and more cost effective would be the easiest solution. But when team member Linus Liang toured a hospital in Nepal, he noticed something strange—the incubators were sitting empty. After interviewing a doctor about this, he learned that many homes where these babies were born were 30 or more miles away on rough rural roads, and

that the parents faced the fight for their babies' lives at home, without much hope of making it to a hospital.

In response to this insight, the design team chose to focus not simply on lowering the cost of incubators but on finding a way to incubate babies at home. The result was Embrace, "a portable incubator, much like a tiny, heated sleeping bag." But the team's work was not done. Design thinking stresses the importance of experimentation and prototyping, and when the team showed their prototype of Embrace to mothers, healthcare workers, and shopkeepers, additional discoveries came to light that would eventually be used to design a device with minimal barriers to adoption. For example, one village mother in India explained that villagers routinely halve doses of Western medicine because they believe the medicine to be too powerful. The original design of Embrace had a temperature indicator strip that could be prone to 'halving' in this way (i.e., to be set at half the recommended temperature, which would be too cold), so the design team removed the temperature strip entirely and instead showcased a simple "OK" indicator.

Design Thinking & Customer Transformation

The design thinking paradigm is sometimes expanded beyond the three phases described above to include five or even six distinct phases. The following six-step version is commonly used:

1. Empathize. Begin by understanding the customers' needs, preferences, and pain points. Conduct interviews, surveys, and other forms of research to gain insights into their experiences and expectations. Sarah Gibbons, Vice President of Nielsen Norman Group, recommends asking yourself questions such as "'what motivates or discourages users?' and 'where do they experience frustration?' The goal is to gather enough observations that you can truly begin to empathize with your users and their perspectives." This deeper empathy for customers is crucial for designing customer-centric solutions that actually work. As Fredierik G. Pferdt, reminds us, "It's great to create an innovative product or service. It's not so great to build one that's useless to people. That's why your users should be your number one focus."

2. *Define.* Based on gathered insights, identify and articulate the most significant customer problems or opportunities for improvement. Develop clear problem statements that focus on the customers' perspective and can be used to guide the ideation process.

3. *Ideate.* This is the heart of the "expansive thinking" aspect of design thinking and centers around creative brainstorming. The goal is to generate a wide range of potential solutions to the defined problems or opportunities by encouraging diverse perspectives and fostering a collaborative environment where team members can build on each other's ideas. Tremendous synergy is possible with such an approach. As Alicia Paterson, a digital futurist and ex-Google transformation leader, commented in a recent interview with me, "the different lived experiences of team members can be a very powerful multiplier in the creative process, as one team member—inspired by the idea of another team member—takes the idea in an entirely new direction the other would likely have never thought of."

One practice I've found particularly helpful for encouraging team members to build upon each other's ideas is to encourage the use of the phrase "yes, and . . ." (which as a side note, is one of the main skills performers learn in "improv.") That is, instead of dismissing or immediately critiquing others' ideas when brainstorming, participants are encouraged to accept and expand upon them by saying "yes, and . . ." followed by their own contribution. This technique fosters a positive, inclusive environment that supports generating diverse and innovative ideas. Paterson recommends, if someone suggests an idea for a new product feature, instead of pointing out potential flaws or limitations, another team member could say, "Yes, and we can also incorporate this additional functionality to address user pain points more effectively." This approach ensures that everyone's input is valued, enabling the team to explore multiple perspectives and possibilities during the ideation phase of design thinking. It's an approach that encourages collaboration, open-mindedness, and creative expansion of thoughts. "It's not just about generating ideas," Paterson noted, "it's also about evolving them."

4. Prototype. This phase involves creating low-fidelity representations of the most favored ideas in order to visualize and test the proposed solutions. Prototypes can be simple sketches, wireframes, or mockups that provide a tangible way to evaluate the potential impact on customer experiences. Indeed, sometimes the simpler and more straightforward the prototype, the better. Alberto Savoia, former engineering director at Google, is a proponent of what he calls the "pretotype," or "a stripped-down version of a product, used to merely validate interest." Building a pretotype has the advantage of "accelerat[ing] learning by building a quick and cheap version of your project" with minimal investment required.

5. Test. During this phase, the goal is to validate the prototypes with actual customers and gather feedback to learn how the customers respond to the proposed solutions. Observe their interactions and ask questions to understand what works and what doesn't. This iterative process helps refine the solution and identify areas for improvement.

6. *Implement.* Once a solution has been validated and refined, you can move forward with implementation. This may involve updating processes, introducing new technologies, or retraining employees to ensure a seamless transition.

Reap the Benefits of Customer-Centric Design

In her article on design thinking, Gibbons points out that "design has historically been an afterthought in the business world, applied only to touch up a product's aesthetics. This topical design application has resulted in corporations creating solutions that fail to meet their customers' real needs." Fortunately, in more recent years, some companies began to catch on to this problem and have "moved their designers from the end of the product-development process, where their contribution is limited, to the beginning." This more customer-centric design approach has "proved to be a differentiator: those companies that used it have reaped the financial benefits of creating products shaped by human needs." In fact, according to the 2015 Design Value Index created by The Design Management Institute and Motive Strategies, "[d]esign-led companies such as Apple, Pepsi, Procter & Gamble and SAP have outperformed the S&P 500 by an extraordinary 211%."

I've surveyed some important perspectives on customer-oriented ideation, which is crucial for carrying out the ongoing process of customer transformation. As we progress, consider the words of Jeanne Liedtka, a faculty member at the University of Virginia's Darden Graduate School of Business: "The most secure source of new ideas that have true competitive advantage, and hence, higher margins, is customers' unarticulated needs.

Customer intimacy [or what we've here called "customer empathy"]—a deep knowledge of customers and their problems—helps to uncover those needs."

Below I've included a table that highlights 10 prominent ideation frameworks, including Design Thinking and 10x Thinking, that businesses can use to foster innovation for their customers.

Framework	Objective	Focus
Design Thinking	Customer Empathy, Creativity, Experimentation, Innovation	New products and services
10x Thinking	Art of the Possible, Large Scale Innovation, Radical Innovation,	Transformative Innovation
Lean Startup	Rapid Experimentation, Iterations, Customer Validation	New Business Products
Agile	Iteration, Product Adaptability, Speed, Collaboration	Software Development
Blue Ocean	New Markets, Creating Differentiation, Low Cost, Business Strategy	Create New Markets
TRIZ	Problem-solving, Patterns, Principles, Creative Thinking	Solve Complex Problems
Six Thinking Hats	Thought Process, Multiple Perspectives, Creativity, Brainstorming, Ideation	Problem Solving and New Ideas
SCAMPER	New Ideas, Modification, Creativity, Innovation, Iterations	Modify Existing Ideas
Model Canvas	Business Models, Business Strategy, Pricing, Market Differentiation	Refine Business Models
Systems Thinking	Problem-solving, Interrelationships, Feedback Loop, System Analysis	Complex Systems

ACTION PLAN

STAGE ONE: CUSTOMER

Introduction

In this first stage, we delve into design ideation. We will explore various ways to empathize with your customers, envisioning what they want and need. This phase is heavy on brainstorming and requires a deep understanding of your customers. The objective is to clearly discern your customers, what they want, and what goals your business should set to cater to them. Remember, this stage is focused on your customers' goals, not your business or technology objectives. Still, the findings in this phase will profoundly influence those later in the process.

Day 0: Reflection and Goal
- **Reflection:** Think about your current understanding and perception of your customers. Consider any assumptions you have about their needs, wants, and aspirations. Evaluate your definition of customer empathy and consider how this understanding currently shapes your business strategies. Recognize areas where your business might not fully meet your customers' goals.

- **Goal:** Your aim for this stage is to cultivate a deeper, more empathetic understanding of your customers and to shift your mindset to a customer-first approach. We want to challenge your preconceived notions about your customers and strive to align your business strategies with their actual needs and aspirations. Before we dive into the ideation process, we aim to set a solid foundation of empathy and understanding, ready to generate transformative ideas for your customers. This mindset will be crucial for the effectiveness of the upcoming sessions.

Day 1: Content Walkthrough
We'll walk through design thinking and ideation concepts and understand how to use imagination and customer empathy in brainstorming sessions. We'll focus on how to develop a customer-first mindset and how this can lead to customer transformation.

Day 1: Workshop Questions
1. How can ideation processes influence the way we perceive our customers' aspirations?
2. In what ways can customer empathy be manifested in our products or services?
3. How can brainstorming lead to innovative solutions that align with our customers' goals?
4. How can we incorporate customer feedback into our decision-making process?
5. How can identifying and understanding our customers' goals change our business strategy?

Day 2-7: Homework and Next Steps
Over the next few days, engage with customers, empathize with them, and brainstorm innovative solutions that align with their needs and wants. Reflect on the questions from Day 1 and identify how your organization can adopt a more customer-centric approach.

Day 8: Touchpoint and Next Steps
1. Review your findings from the brainstorming sessions.
2. Discuss any changes in your perception of your customers and their needs.
3. Identify the top customer goals that align with your business.
4. Develop a preliminary action plan to meet these customer goals.
5. Schedule the next meeting to review and refine this action plan.

Day 9 and Beyond: Action Plan

30-day plan: Cultivate a customer-first mindset within your team. Engage in regular brainstorming sessions and begin integrating the identified customer goals into your strategies. Foster a culture of experimentation. Encourage employees to learn from failures and iterate their ideas. Create a supportive environment where calculated risks are encouraged and groundbreaking projects are supported.

60-day plan: Start implementing the changes based on your new understanding of your customers. Measure the impact of these changes on customer satisfaction.

90-day plan: Evaluate the success of the changes implemented. Refine your approach based on the results and feedback, and repeat this process.

STAGE TWO: INTERFACES

Innovate with People Interfaces

THREE

PEOPLE INTERFACES

"If there's any object in human experience that's a precedent for what a computer should be like, it's a musical instrument: a device where you can explore a huge range of possibilities through an interface that connects your mind and your body, allowing you to be emotionally authentic and expressive."

Jaron Lanier (often considered the "Father of VR")

Not Your Parents' Gas Pump

There was a time when the majority of a business's interactions with its customers were conducted face-to-face. Gone are the days when the customer in question was standing opposite the cash register, seated at a table inside a classic roadside diner, or peering out from a partially opened front door in response to the knock of a traveling salesperson. Today, most business-customer interactions occur through the ubiquitous screen. The device you probably have next to you while reading this book has one (or the screen you're reading this book on). There are likely more screens in the same room you're in, along with the countless other devices we interact with daily: TVs, laptops, smart home controls, self-checkout lines, refrigerators, car dashboards, and gas pumps.

And these screens are growing increasingly interactive. Take the touch screens on gas pumps, for example. In the early days, screen interaction was highly limited, focused primarily on the purchase event: The customer selected a gasoline grade and inserted a card for payment, whereas the pump indicated the amount of gas dispensed, the price per gallon, and the purchase amount. Today, interactivity is much more expansive. One state-of-the-art gas pump from Dover Fueling Solutions boasts a 27" touchscreen display with a welcome screen that recognizes and greets the customer by name, offers real-time weather and traffic information, and provides targeted advertisements and promotions. Customers can even visit the relevant website and personalize their gas pumping experience for items like preferred fuel grade, language, and the information and entertainment options presented at the pump.

We used to call these interfaces customer "touchpoints," but that has become a misnomer: In the age of speech recognition and touchless technologies, we can often get information with either a verbal request (as with Alexa or Siri) or through other means such as gestures, hand interaction, eye tracking, or biometrics such as facial recognition and contactless fingerprints—all without ever "touching" anything.

Their existence dramatically increases the power of such technologies within larger ecosystems of interfaces that make up the Internet of Things (IoT), "the billions of interconnected physical 'devices' or 'things' all around us, exchanging data and communicating between each other over the internet." These digital ecosystems exist in almost every business sector. They involve devices that can be seen everywhere, such as health tracking wearables, smart home devices, residential security systems, self-driving agricultural machinery, intelligent solutions for factories, and technology for monitoring logistics. Notably, of the nearly 22 billion active connected devices worldwide as of 2020, the number of IoT-connected devices (nearly 12 billion) had by the end of that year (for the first time) surpassed the number of connected devices that are not part of the IoT (10 billion). And this number of IoT-connected devices is growing exponentially: a 2021 report estimated there will be nearly 56 billion such devices by 2025.

This interconnectedness opens up possibilities for how businesses and customers interact. Back to the gas pump, imagine when (likely very soon) the ecosystem of interfaces grows to allow your car to communicate directly with the gas pumps at stations on your route ahead. This will make it possible, for the nearest pump to expect your arrival and know precisely how much gasoline your vehicle will require. We are already seeing something similar with current EV route planning apps, which direct drivers of electric vehicles to the nearest available EV charging station, in some cases taking into account current road conditions and even which route will be least taxing on the car's battery life.

Another game-changer for creating highly interactive, personalized connections with customers is the advent of artificial intelligence (AI) as a pervasive element of the digitalization of businesses. A great example is the beauty line Ulta, which has built virtual experiences for their customers based on AI-generated data, from a "Virtual Try On" session that promises true-to-life textures, colors, and finishes to a "Complexion Matching" feature that detects the customer's skin tone and undertone to find their best product matches. Ulta even offers an app that can recommend a personalized skin-care routine through the use of facial recognition technology.

Connecting People to People

What are the implications of these rapid technological advances affecting how companies interact with customers? In the first two chapters, I emphasized the importance of allowing your business decisions to be informed and guided by a deep knowledge of the ever-evolving needs of your customers—a process I call *customer transformation*, which is based on a mindset often referred to as the *outside-in perspective*. Against this backdrop, let's explore two essential points regarding the explosion of digital interfaces in today's marketplace.

First, recall my point in Chapter 1 regarding technicians, marketers, and executives often focusing on how digital transformation will upgrade or otherwise add value to the business rather than how it will better meet the customers' needs. In actuality, these same decision-makers often view digital interfaces—whether between the company and its customers, the company and other companies (partners, developers, other service providers), or even between the employees within the company itself—as merely linking one tech application to another rather than fulfilling their far more critical function: *linking people (through technology) to people*.

The importance of this point first hit me a few years ago when I was doing work in the API space. I had long believed strong businesses link every decision back to improving their value propositions. Through my work with APIs, I discovered that this tech delivers the most value when it serves as an interface between a company and *people*. I noticed that when the development of an API program was driven solely by the desire to link applications to each other, the result was often a one-size-fits-all API built with the company's internal infrastructure in mind and largely disregarding the needs of customers or other users. But when an API program was viewed as a link between people and technology, rather than merely linking technology to other technology, then every proprietary API the program created and every third-party API it leveraged could potentially drive incremental value for the business.

Thus, when coaching enterprise leaders set to launch an API program, I often began by asking, "What communities and individuals do we want to connect with our digital assets?" A business might want to penetrate the smart home market, in which case it would be crucial to recognize that the developers building smart home applications are not necessarily the same developers who build typical smartphone apps. They may be motivated by specific technologies and financial incentives not shared by other developer groups. This would likely require API design considerations unique to the smart home market. APIs may be the universal connector between applications, but that doesn't mean all APIs should be universally the same, precisely because the people at the ends of those applications are unique.

For this reason, I've titled this chapter "People Interfaces." Despite the sophistication of devices, screens, or other touchpoints (or "touchless" points) involved, these points of interaction acquire their significance from the people who employ their use (a truth we often ignore at our own peril). It's quite easy to get caught up in technological wonder and forget this human element—until you become one of many who admit to saying "thank you" to their home devices after a response. We have a craving, it seems, to find human connection in even our most technical endeavors.

Tech that Enables Authentic Moments of Action

My second point relates to how these interfaces—technological points of interaction between people—relate to customers or other users' realities, especially as these realities are embedded within or affected by the rapidly expanding, interconnected digital ecosystems mentioned earlier. It is crucial to see these points of interaction as defined not by the tech itself but by the customer's perception of the exchange (i.e., the *reality* they experience in that moment of engagement with the company through the tech) and by the actions the customer may be motivated and enabled to take within that shared reality. Commitment to customer transformation establishes access to a data ecosystem capable of innovating, transforming, and powering reality systems in which your customers experience the physical or virtual world in new ways that create pivotal moments for their actions.

As companies continue to invest in people interfaces, it is crucial to remember that these interfaces are only one aspect of the larger digital ecosystem in which customers and users exist. It is not enough to offer a chatbot or voice assistant—companies must also consider the overall customer experience and how it fits into the broader context of the customer's reality. This means considering the customer's location, personal preferences (like dietary or accessibility), behaviors, and other technologies and platforms they may use. By understanding the customer's reality and enabling transformative experiences, companies can create more meaningful customer connections and build long term loyalty.

Looking ahead, we can expect to see even more advanced and sophisticated people interfaces in the near future. Multi-sensory interfaces are found in various applications, from gaming and entertainment to education and healthcare. For instance, gaming applications often use haptic feedback and audiovisual effects to create a more immersive player experience. In education, multi-sensory interfaces provide more engaging and interactive learning experiences for students.

One of the most exciting developments in multisensory interfaces is the incorporation of virtual reality (VR) and augmented reality (AR) technology. With VR and AR, users can experience a highly immersive and interactive environment that engages every sense, from sight and sound to touch and smell. This technology has a wide range of applications, from gaming and entertainment to education, healthcare, and even mental therapy. Imagine a device that allows you to be anywhere in the world through VR visually but also provides the scents of that location—users could not only experience the Bahamas visually, but also smell the salt and distant aromas of hibiscus and bananas.

In 1960, the *Smell-O-Vision* system was created to emit scents synchronously with the projection of a film, allowing theater viewers to experience aromas that corresponded with each scene. Developed by Hans Laube, the technology was featured exclusively in the motion picture *Scent of Mystery* produced by Mike Todd Jr., the son of a renowned film producer.

The mechanism involved the release of 30 distinct fragrances that were triggered by specific cues in the film's soundtrack and emitted from movie theater seats and. In essence, *Smell-O-Vision* offered a multi-sensory experience, allowing the audience to see, hear, and smell what was taking place on screen. This type of multi-sensory interface, however, wasn't the first attempt to introduce smell into the theater experience. The concept dates back to 1868 for live performances, then again in 1953 with General Electric's *Smell-O-Rama* and its competition *AromaRama* in 1959.

Disneyland Resort offers similar experiences in a few of its rides. Soarin' takes travelers to various locations worldwide, providing scents throughout the journey: pine trees in Redwood Creek, oranges from Valencia Orange Farms, rose blossoms at the Taj Mahal, and a sea breeze in the South Pacific. Perhaps Disneyland's best experience is during the holiday season when the park transforms *The Haunted Mansion* for Halloween, complete with a haunted house-style gingerbread house that surrounds your "doom buggy" with a gingerbread scent.

These technologies can create immersive interfaces that blur the lines between physical and digital reality, allowing customers to interact with products and services in entirely new ways.

In 2021, Japanese professor Homei Miyashita, developed a prototype lickable TV screen that simulates food flavors. The device, called Taste the TV, houses a carousel of flavor canisters that spray a combination of particles to replicate the taste of a particular food. "The goal is to make it possible for people to experience something like eating at a restaurant on the other side of the world, even while staying at home," the professor said. This technology has the capability to transform the way we perceive food and may have practical uses in domains like cooking, entertainment, and healthcare. In critical environments such as during the COVID-19 pandemic, this technology could improve how individuals connect and engage with the external environment.

A Lab in the Palm of Your Hand

The single most used interface in the last several years has been the "COVID test." The 2020 pandemic upended society in numerous ways, one being that it exposed shortcomings in systems such as healthcare to which most of us had grown accustomed. We heard a great deal about the need for "rapid" and "accessible" testing during the pandemic, but the reality, for many, was far from it. Analyzing the testing process's moment-by-moment user experience, it's not hard to discover frustration at every step.

One startup, Cue Health, recognized that the Covid test was an example of traditional tech that badly needed revamping in light of the customer's experienced reality. As one of Cue Health's founders said in a podcast interview with me in December of 2021, "It shouldn't be hard . . . to get that information . . . Being able to go from feeling symptoms to getting a diagnostic test to being able to connect that to a doctor and then to get the treatment delivered to you in a very short amount of time—that seems like the ideal pathway for which all infectious diseases should be handled."

This startup's solution was Cue, a handheld diagnostic testing kit with a built-in "lab" component that provides rapid results in minutes through the Cue Health app on the user's phone. The app is connected to the cloud, so in addition to receiving truly "rapid" results, the user can immediately coordinate virtual doctor care and receive a prescription, which can then, in many cases, be filled and delivered the same day to the customer's home. The genius of this approach is that it radically recreates the consumer's moment-by-moment experience of interacting with healthcare providers/services to receive a diagnosis and treatment in at least two ways.

First, the process is dramatically reduced from what was more than a week in the traditional scenario to just a day or two in the Cue scenario (the only significant wait times being the day it may take to receive a Cue test in the mail and perhaps the additional day to receive one's prescription). Second, and equally important, the Cue scenario offers greater actionable control to the customer, who can easily administer their diagnostic test and directly control how the results are used. This contrasts with the all-too-common feeling of being at the mercy of multiple individuals (medical technicians, assistants, nurses, doctors, pharmacists) regarding the speed and ease with which the needed diagnosis and treatment is received. In short, Cue Health has created an innovative data ecosystem that "transforms and powers a new system of reality" for the customer—namely, their diagnostic testing experience.

A Zebra Who Changes Its Stripes

Several years ago, Zebra Technologies, the leader in barcode scanning technology, asked me to conduct a digital strategy workshop. The workshop was the result of the company's decision to transform from a hardware company to a software-based company. This decision was motivated by a simple observation: devices in people's hands kept changing, but essentially, the same software could be used to scan barcodes through any number of devices at any time and anywhere.

On the first day of the workshop, which included about 25 people with various roles and responsibilities, we went through numerous decks, Q&A sessions, and hypotheses. It was educational, but we were not progressing the way the participants expected. By the second day, the number of attendees had decreased to about 15. Despite our best efforts to discover ways of making Zebra's data more meaningful, the workshop did not have the impact everyone hoped for.

I entered the conference room on the morning of the third day and could sense the frustration. There were now only 6 participants, and it seemed these individuals were simply going through the motions. Knowing that we needed to change our approach dramatically, I candidly told those present that I was scrapping the original agenda and asked them to gather as many people back into the room as possible. I wanted to try something new.

Fifteen minutes later, I stood before the re-assembled group and announced, "Let's look at this issue differently today." The skepticism was plain on their faces. I cleared the whiteboard and broke them into small groups to brainstorm new perspectives. I recommended an outside-in approach and prompted them to consider how the latest barcode scanning software would interface with people and how they might make the data more meaningful to the customer. I remember at one point saying, "Think about the exact moments that people have engaged with your hardware in the past, and consider what will happen in the same moments once you've converted the capacities of that hardware into software—into monetizable data points."

In the seconds that followed, there was what we consultants all love to hear: silence.

Then, slowly but surely, people began to nod and murmur between each other. The room became louder as breakout side conversations sprang up. Minutes later, one of the directors in the meeting turned to me and asked in front of the others, "Would it be possible for us to develop an entire ecosystem of this data between our business units, partners, and customers that they could interface with?" The room paused, awaiting my response.

"Absolutely," I said, and the room sprang to life again. It was their "ah ha!" moment.

Excited and energized, the participants continued to work through the rest of the day, building on the breakthrough we had just experienced. By the end of the workshop, we had developed a clear plan for how Zebra Technologies could make its data more meaningful and shift its mindset from (a) manufacturers of a handheld hardware device designed to perform a function to (b) providers of a ubiquitous people interface that delivers authentic, data-driven, actional moments for the user engaged with the technology.

Real Solutions

As technology evolves at an unprecedented rate, it's easy for businesses to become preoccupied with incorporating the latest and most remarkable technologies into their operations. However, it's important to remember that technology is ultimately just a means to an end and that the most critical consideration for any business should always be the customer experience.

Rather than focusing solely on how your business will interface with technology, it's vital to consider how customers will interface with you. Each of us encounters a multitude of screens and devices daily that inundate us with sensory-driven interfaces. We engage through countless interactions with technology that form crucial moments in our experience of reality and pivotally affect our actions. To be able to not only recognize such moments of action in our customers' lives, but deliver real solutions within these moments—solutions that in some way significantly improve the quality of customers' moment-to-moment experience of the world—is the purpose of people interfaces.

FOUR

ARTIFICIAL CUSTOMER RELATIONSHIPS

"A computer would deserve to be called intelligent if it could deceive a human into believing that it was human."

Alan Turing (1950)

Are We There Yet?

Back in 2014, there was some fuss over a report that 64 years after Alan Turing presented his famous Turing Test—a simple indicator of when a computer would "deserve to be called intelligent"—one had finally achieved this milestone. There is still disagreement over whether the computer in the 2014 case (which involved "fooling several judges by successfully impersonating a 13-year-old boy") or the ones since that time should count as having passed Turing's test as he intended it. However, no one can deny that Artificial Intelligence (AI) is beginning to have a transformative effect across industries as well as societies. And there appears to be no end in sight for AI's influence: As David Vandegrift, CTO and co-founder of the customer relationship management firm 4 Degrees, cautions, "I think anybody making assumptions about the capabilities of intelligent software capping out at some point [is] mistaken."

In chapter 3 I mentioned that AI is a game-changer for creating highly interactive, personalized connections with customers. As such, AI has immense potential for advancing customer transformation. But there is a caveat: Just as AI functions as an "artificial" form of intelligence, the *relationships* AI attempts to forge with customers are "artificial" as well.

In this context, AI isn't necessarily a bad thing. AI's artificiality is its ability to perform complicated tasks and achieve outcomes that no human could ever pull off, which is why it brings such value to the table. AI can sift through mountains of data to provide businesses with a more profound understanding of their customers' needs and make their interactions more efficient, productive, and, potentially, more satisfying.

Given this knowledge, we must seriously consider whether our use of AI will positively or negatively affect the business-customer relationship. The best way to predict this is to determine how closely our use of AI aligns with the customer's specific, fundamental values. To the extent that our use of AI supports these values, our customer relationships strengthen; when we ignore these values, our use of AI risks alienating our customers.

However, just as the emphasis on digital transformation can distract us from the fundamental importance of customer transformation, an unmitigated focus on AI can quickly counter our attempts at genuine awareness and empathy toward customers. The potential loss of human connection due to the presence of AI runs the risk of generating artificial relationships.

Customer Values
Privacy

AI's perceived threat to privacy has received much attention in the media as well as academic and business circles due to the ethical concerns involved. One AP story shares Walgreens pilot program in some stores to connect cameras and sensors to beverage coolers. "Instead of the usual clear glass doors that allow customers to see inside, there are video screens that display ads along with the cooler's contents." Above the door handle is an AI-connected camera that interprets age and tracks eye movement.

Despite what Pam Dixon of the World Privacy Forum calls the "creepy factor" surrounding surveillance, advocates of this technology argue that it allows businesses to display "discounts tailored to" the particular customer—what would seem to be a customer-centric objective. These same advocates often dismiss privacy concerns by pointing out that, in many cases, "the information collected is anonymous." However, critics like Ryan Calo, a professor at the University of Washington School of Law and co-director of its Tech Policy Lab, note that even anonymized data is used in intrusive ways:

> For instance, if many people are eyeing a not-so-healthy dessert but not buying it, a store could place it at the checkout line, so you see it again, and "maybe your willpower breaks down. . . . Just because a company doesn't know exactly who you are, doesn't mean they can't do things that will harm you.

Put differently, just being customer-centric isn't enough when striving to understand the customer. Whether a business uses the information collected via AI to benefit the customer also matters. A critical reason that privacy is considered a "right" of individuals in the first place is precisely that their protection tends to yield significant value for those individuals. A complete perspective on customer transformation keeps this focus on the benefit to the customer front and center, which in turn means aligning AI and other technologies with the customer's values—such as privacy.

Equity

Another customer value that AI can potentially undermine (with the corresponding ethical concerns this brings) is equity or our sense of fairness. Stories like Walgreens willingness to consider using AI to track the age of individual consumers and offer targeted advertisements on this basis open the door to privacy and equity concerns, given the fine line between targeted advertising and discriminatory practices. As Guha et al. point out in their article, *How Artificial Intelligence Will Affect the Future of Retailing*, "even if race or gender is not a formal input into an AI algorithm, an AI application may impute race/gender from other data and use this to price higher to specific demographics." Marketing expert Ron Ventura agrees:

> [L]everaging real-time and rich data on differences between customers (e.g., in preferences and price sensitivity) can lead to increased discrimination and greater social inequality among consumers. The emergence of AI is likely to change the nature or form of relationships and relationship marketing: it may create new possibilities but also raise or [emphasize] problematic issues; AI may render relationship marketing more accurate and scalable, yet also more discriminating.

Along similar lines, researchers Cui-Mertz and Jung note that retailers increasingly use perfectly legal AI applications to make "purchasing recommendations based on facial monitoring and tracking [the] mood of in-store customers." Again, the line here between what is appropriate and what does not, seem tenuous: Imagine the implications of retailers pushing certain medications to "sad" customers, for instance.

Some may claim that concerns about customer values like privacy and equity are only worth considering when the customer is aware of their existence and consciously objects to them. Others may believe there is no cause for concern if the customer is aware of the presence of AI tools and isn't concerned. For example, one New York Walgreens shopper told a reporter testing cameras and sensors similar to the ones described earlier that he was unconcerned since such cameras are practically unavoidable. "There's one on each corner," the shopper said dismissively.

This perceived lack of concern could be interpreted as consumers waiving their rights to privacy and equity. But it would be a mistake to conclude that the possibility of such waivers means businesses have the latitude to ignore customer values—if for no other reason than customer reactions to technological intrusions are not entirely predictable, and customer backlash is an ever-present danger in the marketplace. Indeed, in the same New York Walgreens mentioned above, another shopper clarified their objection to surveillance cameras. When the reporter pointed out the camera, the shopper responded, "I don't like that at all."

Autonomy

Another customer value that may be the most important to focus on when considering how productive and responsible use of AI can strengthen a business's customer connections: autonomy, or the customer's sense of having control over their own decisions and actions. One reason this value is so significant is that even when other values appear threatened, the customer's perception of lost autonomy often fuels backlash and general discontent with the business. Ventura touches on this when summarizing what can happen when the use of AI leads to discriminatory treatment of customers and thus violates the customer's sense of equity (based on Libai et al.'s research):

> Customers who are either less financially capable to make large expenses or less willing to be committed to a given company may find themselves in a greater disadvantage in their relationship with the company compared with higher value customers [i.e., due to differential treatment from AI], and they . . . could face another consequence: *feeling that they lose some of their autonomy,* which . . . may increase the sense . . . that they are being manipulated or discriminated against.

Customers' concerns over the invasion of their privacy can be similarly connected to a perceived loss of personal autonomy. In more severe instances, there could be unintentional leaks of sensitive data from the system to others. Consumers might also feel as if they're under constant observation by cameras or more susceptible to loss of their rights when they entrust intelligent assistants with purchasing decisions or acting on their behalf. Such occurrences raise concerns over losing personal control and prompt consumers to demand greater transparency regarding how their data is used.

But the use of AI can potentially undermine the customer's sense of control beyond its effect on other values; AI may impinge on customer autonomy in more direct or, sometimes, more subtle ways. One context in which this may occur is when AI algorithms are trained to respond to customers' shopping habits. Ventura explains:

> Libai et al. highlight a shift in focus from customer loyalty to habits. Customers' habits are acknowledged as more characteristic of consumer behaviour [sic], a driver of market success, and [a factor] which AI can more easily draw on and leverage. (Note: this approach no longer requires the establishment of attitudinal loyalty, just the prevalence of strong habits.) Artificial intelligence can help fulfill two functions in this regard: (1) "teach" and form new habits for customers to practise, and (2) reinforce, reward, and enhance existing habits.

This focus on customers' habitual behaviors can bring clear benefits to the company and the customer. For example, because "[h]abit forming . . . is governed by automaticity and requires minimal cognitive attention for performing frequently repeated behavior," AI-driven shopping recommendations may cause a customer to feel "more satisfied with the outcomes and happy with the less effort needed on his or her part to make purchase decisions," in which case "he or she will be more likely to adopt this way of purchasing as a regular habit." Establishing such habits drives up switching costs and thus allows companies to better "protect their customers from churning."

On the other hand, a mindset that sees the customers' value to a business reduced to the sum of their reinforceable (and manipulable) habitual behaviors—and that views AI primarily as a highly sophisticated precision tool for accomplishing this reinforcement—is in tension with a customer transformation mindset. The mantra of customer transformation is that *customers must always come first*—not just in what they are demanding (an efficient technical interface, a frictionless customer experience) but in who they are as people that engage with your businesses and in what they need at the most fundamental human level. It's important to recognize humans' central need to be valued as thinking, feeling, and autonomous individuals. Therefore, a sustainable approach to customer transformation must include genuine empathy toward customers and respect for their autonomy.

Another use of AI that may end up diminishing customer autonomy is, ironically, when it's offered as a self-service resource (i.e., to accomplish consumer-related tasks without the need for human customer service representative assistance). Generally speaking, this use of AI is designed to increase consumers' autonomy, not diminish it, by allowing consumers to do more unassisted. Many such experiences are instances of the much-discussed concept of "frictionless" interactions, or what Anthony Smith calls the "lean back" experience,

> in which a customer lets the machine or service do the dirty work. The user puts in as little effort as necessary to achieve their end result. Think of Amazon—not only is a product delivered to you directly, sometimes in a matter of hours, but you barely even need to search for the product in the first place. The more you use Amazon to search for products, the more signals you generate, the more it knows what you're looking

for, and ultimately the better it can provide personalized suggestions along the way. Friction is reduced at every turn.

When a customer can rely on AI to "do the dirty work" without any help from a human customer service representative, this would seem to align with the customer's value of personal autonomy. In the words of Castillo et al., "[b]y interacting with AI technologies to self-serve, a customer becomes a central element of service production, a 'partial employee' and a co-creator of value." And as Smith points out (citing Gartner), "next-generation customers strongly desire self-service. These customers don't want you to put humans in their way if they could complete a task independently. They don't see the need for another person to be involved when technology so rarely fails them.'"

The problem arises when the customer is asked to rely on AI but either lacks the know-how or resources to co-create value with the AI (e.g., doesn't understand how to interact effectively or overestimates the AI's capabilities.) Castillo et al. explains:

> In the same way, that value is collaboratively co-created, it can be collaboratively co-destroyed during the process of interaction. AI technologies rely on customer participation, which increases service complexity and, eventually, the likelihood of service failure. As customers invest higher levels of effort and time into an interaction, they might feel annoyed and frustrated when the co-created service fails to meet their expectations. Indeed, these instances represent the loss of valuable resources, such as time and patience, for the customer.

In such cases, customers' perceived autonomy is threatened because they cannot exercise sufficient control and achieve desired outcomes (e.g., a quick, convenient transaction).

This chain of events leading to unmet expectations is an example of "resource misintegration," which occurs when one of the agents in a business-customer interaction lacks knowledge or other resources to help initiate a successful outcome. The customer lacked the needed resources in the preceding case, but the business's AI agent can also fall short. Think back, to a time when you've interacted with a chatbot that was unable to comprehend or accurately respond to your specific request. If you became frustrated or even angry, it was most likely because you sensed your autonomy over the situation threatened:

> Although customers may not expect the chatbot to fully resolve their problems or issues, they do expect that, at a minimum, it is able to understand the context of their question and to provide adequate guidance. When this is not the case—for example, when the chat results in a deadlock—customers feel agitated and upset because they feel that they have lost control of the interaction.

This "deadlock" can leave the customer feeling incapable of moving the interaction toward a satisfactory resolution and cause them to lose trust in the AI's service provider rapidly. Indeed, Castillo et al.'s research found that customers tend to take a dim view of such outcomes: "Resource misintegration is perceived as an intentional action by service providers to maximize benefits for themselves." There's certainly no faster way to drive a nail into the coffin of the "human connection" we are hoping to nurture.

Respect

The fourth and final value with which AI uses should align is closely associated with the previously discussed customer autonomy but is broader and deserves special mention. Customers need and expect to be treated with respect by those they transact with, meaning they want to be treated like persons with legitimate needs, wants, expectations, abilities, rights, and feelings. Although customers are indeed looking to maximize the efficiency of their transactions with businesses, this doesn't mean they want those transactions stripped of all traces of humanity. And if existing research is correct, it may not be easy to compensate for this by having AI become more and more human-like: As Castillo et al. points out,

> [s]tudies of human-computer interaction have reported that when chatbots are designed to be more complex and animated, exhibiting high levels of anthropomorphism, customers experience the negative feelings of eeriness and unease (Ciechanowski et al., 2019). . . . [which] can lead to negative attitudes towards a service provider with resultant—and often irreversible—reduced purchase intentions (Demoulin & Willems, 2019).

The negative impact was worsened if the customer suspected that the identity of the chatbot had been "shrouded" from them: "Customers felt deceived when they were made to believe they were interacting with a human agent but found out that they were interacting with a chatbot."

The upshot is that it would be unwise for businesses to become so enamored of AI's capabilities that they assume AI can entirely replace human agents. Statistics strongly support the need for caution in this regard: According to a PWC study, 75 percent of consumers indicated that they will still choose to interact with a real person even as the technology for automated solutions improves, and "[o]nly 3% of U.S. consumers . . . want their experiences to be as automated as possible." Moreover, far more Americans strongly disagree (55%) with the following statement than strongly agree (23%): "Once the technology becomes advanced, we won't need people for great customer experiences."

The resource misintegration scenario described earlier is a case in point: Puthiyamadam and Reyes pointed out that when people engage with apps, self-service checkouts, websites, and the like and "something goes wrong, they want to talk to a person, stat." The worst thing we can do—from a customer transformation perspective—is structure our interactions such that customers may find themselves "caught in a process" and trapped in a room with AI . . . no authentically human exit in sight.

Ways to Maintain the Human Connection
The Connected CIO

I remember conversing with the CIO of an extensive product manufacturing company during a workshop I was leading. We discussed their current use of AI in their sales team's inventory and sales calls processes. The CIO was excited that the company was building a new AI tool to update its sales team about inventory and sales calls through a mobile app.

Knowing this use case was already outdated, I suggested he consider how that same process could be automated so that his field and sales reps wouldn't need to make the calls or visit the stores. Using predictive analytics and machine learning, I explained that supply levels at retail locations could automatically be calculated based on purchase history. The AI would submit a new order when appropriate and have the order fulfilled without any human contact. This would streamline the process, reduce the chances of error, and optimize the supply chain.

"But we'd lose the human connection," the CIO replied, skeptical of my proposal.

I paused and started to respond but stopped myself. Honestly, he had a point and at that moment I wasn't sure how to reply. Perhaps I had let my eagerness to apply AI's wonders eclipse the importance of human connection. I suggested we continue this discussion in tomorrow's session, and the CIO graciously obliged.

Sitting in my hotel room that evening, flipping through television channels, I stopped on a "Star Trek: The Next Generation" episode titled "*The Measure of a Man.*" The episode explored whether Data, an android played by Brent Spiner, was a legally sentient being with the full range of rights and freedoms. Data's character represented an artificial intelligence striving to be more 'human' and, thus, more relatable.

By the end of the episode, I realized that much of the episode could be applied to businesses leveraging AI to build customer relationships. Although AI can undoubtedly provide highly efficient, personalized interactions, current AI lacks 'the human element' needed to make customer interactions more meaningful and memorable. Although there is indeed the "eeriness" factor to consider, there would seem to be room for businesses to create a more empathetic and relatable customer experience by seeking to humanize AI.

Data's journey to become more human highlights the importance of understanding and respecting our values, emotions, and experiences. When developing AI for customer relationships, it is especially relevant given the concerns surrounding the ethical and social implications of AI in customer service. Just as Data learns to navigate the complexities of human emotions, businesses must also strive to ensure that AI is developed in a way that aligns with human values and emotions.

The CIO had a point: human connection is vital. But I also knew that AI could enhance this connection rather than replace it. The next day, during a break, I pulled the CIO aside and thanked him for his patience, telling him I had given his comment from the day before a great deal of thought. After relaying some of my Star Trek-inspired insights, I supplemented my previous proposal with a suggestion: "Have your sales reps used the extra time freed up by automation to focus on building stronger relationships with customers? The sales reps could leverage the data generated by the AI system to make more informed and personalized recommendations for customers. Rather than removing the human connection, this would enhance and prioritize it.

The CIO realized the value of my suggestion and nodded in agreement.

Two Principles

So, how can we use AI to support rather than diminish the human connection with customers? Let me suggest two basic principles, each with many applications.

1. Strike a balance. As I suggested to the CIO that his company reap the efficiency benefits of AI but supplement AI's usage with a renewed commitment by his sales reps to customer relationships, be sure to complement your use of AI with authentic human connections. In his Forbes article, Anthony Smith makes essentially this same point about conversational chatbots, which, he suggests, "[w]orking alongside human employees, . . . can serve as frontline engagement tools that help drive customer interactions and funnel prospects to the appropriate channel where a human resource can truly add value." On the one hand, Smith observes, AI's "[c]onversational interface capacities provide personalized experiences that cull and sort information faster than humanly possible." On the other hand, this "greater operational efficiency . . . frees up human resources to truly impact and delight customers and build brand loyalty."

Emily Chantiri describes just such a successfully balanced use of AI by the online consumer giant Alibaba, whose AI-bots

> continuously monitor whether customers encounter obstacles and whether the AI-based service can understand and resolve customers' queries. Whenever appropriate, the AI bot automatically transfers customers to human-based service agents, who will then prompt and provide agents with essential information to help continue the conversation without asking customers to describe the problem repeatedly.

This balance between AI and human engagement with customers should be mapped out by businesses, much like supply and demand. As the graph on the next page illustrates, with increased use of AI, the speed and overall efficiency of service tend to increase, but the quality of service that can come from the human touch decreases. A balance of positive AI implementation and happy customers is at the intersection—your task is to find that sweet spot for your company and customers.

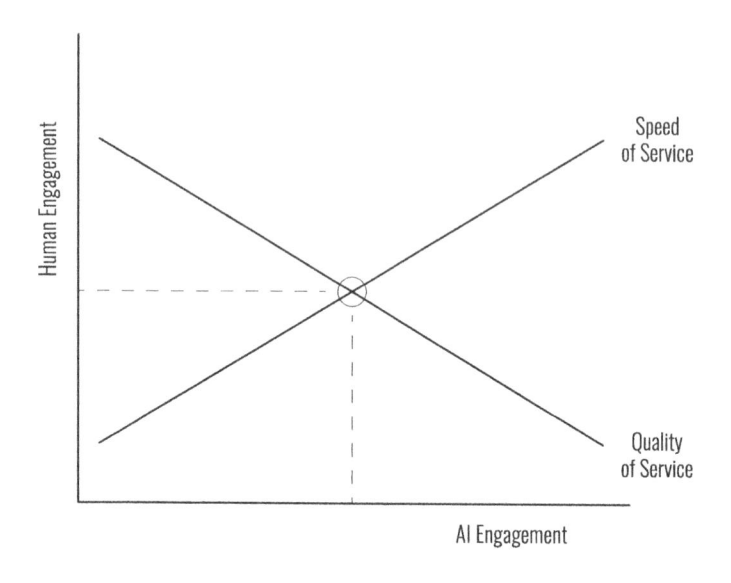

```
                                    Speed
                                    of Service



Human Engagement




                                    Quality
                                    of Service


              AI Engagement
```

2. Focus on the customer's human values. Although customers want transactional needs (such as speed of service) to be met, the empathy we exercise toward our customers must also consider more fundamental human values and emotions, including the three values of privacy, equity, and autonomy. Aligning our use of AI with these core values constitutes a significant part of providing the "quality of service" referenced in the graph above.

Take AI's ability to increase the personalization of services that companies can offer their customers. In an article on developing customer loyalty, Emerson Sklar notes that despite its importance, "75% of consumers state they find certain forms of personalization creepy" and that "[o]ften brands might be doing more harm than good with some of their personalization tactics, which can severely damage their reputation and customer relationships." Sklar advises:

Before retailers consider implementing a personalization strategy, they must not only strive to understand the affinities of the end user but also understand that there is a human being at the receiving end of all this effort. Companies can have the best AI system in place, but if the personalization strategy is not human-centric, and as a result efforts misfire, they may find themselves in a worse position than if they had not implemented the system at all.

Considering your customers' fundamental human values in this way may lead you to hold off implementing specific AI strategies or limit the ability to provide a "frictionless" customer experience. In a 2022 Harvard Business Review article, Renée Richardson Gosline proposes needing at least some "good friction."

Good friction is a touch point along the journey to a goal that gives humans the agency and autonomy to improve choice rather than automating the humans out of decision-making. This decidedly human-first approach allows for reasonable consideration of choices for the consumer and testing of options by the management team according to user needs, as well as a clear understanding of the implications of choices. And it may also enhance the customer journey by engaging users in increased deliberation or better co-creation of experiences.

Respecting your customers' right to privacy, requires that your company introduce good friction by being transparent about its use of AI, even though doing so may make the interaction somewhat less seamless than it would otherwise be. But as Gosline reminds us, "A confident, responsible brand should not have to engage in sleight of hand or manipulation to boost engagement."

Everyone wins

AI is increasingly becoming the core component behind the explosion of new visual and voice interfaces between businesses and customers in the twenty-first century. These interfaces are creating more personalized, automated, and accelerated customer experiences than ever before. But at what cost?

One of the essential insights of a customer transformation mindset is that generating value for a company depends directly on developing value for the customer—not in lip service only, but genuinely so. The point is when considering what customers value, we must include what gives meaning to their relationships and helps form their sense of identity as customers as well as human beings. Suppose we can find ways to employ AI to increase efficiency and productivity and align with and support these fundamental human values. In that case, our application of AI's remarkable capabilities will form an integral part of a successful customer transformation strategy.

ACTION PLAN

STAGE TWO: INTERFACES

Introduction
The second stage is about the evolution from simple "touchpoints" to comprehensive "interfaces" that your customers use to engage with your business. This includes using artificial intelligence, which is rapidly becoming a vital interface in the customer-company relationship. These interfaces, or "People Interfaces," strengthen customer relationships and facilitate the business transformation to continually meet customer engagement preferences.

Day 0: Reflection and Goal
- **Reflection:** Reflect on the current interfaces that your customers use to engage with your business. These could range from traditional mobile and web interfaces to more advanced touchless interactions, voice recognition, AI-driven platforms, and virtual reality experiences. Consider the role these innovative technologies currently play within your interfaces, and consider their accessibility and efficiency in meeting your customers' needs and preferences.

- **Goal:** Your primary objective for this stage is to evolve beyond traditional touchpoints to comprehensive, technologically advanced interfaces, possibly encompassing touchless interactions, voice recognition, AI, and virtual reality. The goal is to harness these innovations to enhance customer engagement and facilitate a transformational customer-company relationship. As we prepare to delve into the intricacies of People Interfaces, your goal is to start with a clear understanding of your current technological footprint and an ambition to innovate. This mindset will lay the groundwork for creating next-generation interfaces that redefine how we connect with your customers.

Day 1: Content Walkthrough
We'll dive into the "People Interfaces" concept and how AI and ML influence customer relationships. We'll also explore the role of APIs, connected devices, voice recognition, seamless integrations, and accessibility in enhancing customer interaction.

Day 1: Workshop Questions
1. How can People Interfaces enhance customer engagement and relationships?
2. What role does AI play in improving customer interfaces?
3. How can APIs and connected devices improve how customers interface with our business?
4. What potential benefits might voice recognition bring to our business?
5. Why is accessibility important when considering customer interfaces, and how can we ensure it?

Day 2-7: Homework and Next Steps
Over the following days, observe and analyze the effectiveness of your current interfaces. Evaluate their accessibility and how they contribute to building strong customer relationships. Start brainstorming how to integrate AI, ML, APIs, and other modern technologies into your interfaces.

Day 8: Touchpoint and Next Steps
1. Share your observations about your current customer interfaces.
2. Discuss how AI, ML, and other technologies can enhance these interfaces.
3. Identify the critical steps needed to improve the accessibility of your interfaces.
4. Develop an action plan to implement these improvements.
5. Schedule the next meeting to review and refine this action plan.

Day 9 and Beyond: Action Plan
30-day plan: Work on enhancing the accessibility and functionality of your customer interfaces. Incorporate customer feedback and new technologies like AI and ML into your interfaces. Map your API program to accelerate products and services for your customers.

60-day plan: Begin testing and implementing the changes to your interfaces. Measure customer responses and adjust accordingly.

90-day plan: Evaluate the effectiveness of the changes and the impact on customer relationships. Refine your approach based on results and feedback, and prepare for the next training phase.

STAGE THREE: JOURNEYS

Engage in the Moment

FIVE

THE OUTSIDE-IN PERSPECTIVE

"You've got to start with the customer experience and work backward to the technology."

Steve Jobs

Value for . . . Whom?

A headline on Variety.com (2023) announced the news: *"Disney+ Drops 2.4 Million Subscribers in First Loss; Bob Iger Heralds 'Significant Transformation' Underway."* The loss occurred in the final quarter of 2022 and marked the streaming service's first decline since its 2019 launch. This report received extra attention because it was "Bob Iger's first quarter back as CEO, after Disney's board ousted Bob Chapek in November, with Iger seeking to reassure investors that the company has a plan to get back on track."

When announcing the quarterly results, Iger delivers a master-class performance in reassurance. In his words:

> After a solid first quarter, we are embarking on a significant transformation, one that will maximize the

potential of our world-class creative teams and our unparalleled brands and franchises. We believe the work we are doing to reshape our company around creativity, while reducing expenses, will lead to sustained growth and profitability for our streaming business, better position us to weather future disruption and global economic challenges, and deliver value for our shareholders.

While a finely crafted message, every element of Iger's transformation strategy, from unlocking "the potential of our world-class creative teams" to "reducing expenses" and delivering "value for our shareholders," is focused on what Disney is going to change or do *internally*. Not once does Iger directly address the customers or focus on what value Disney will deliver to attract those 2.4 million people back to the platform. Iger is so focused on an internal transformation that he completely loses sight of the proposed transformation's touchstone: *the customers*.

A Matter of Perspective

A customer transformation mindset begins with the belief that customers' needs and expectations should be the guiding force behind a company's strategic transformation. This conviction has been termed the "outside-in" perspective. In the words of Harvard Business School's Ranjay Gulati, who originated the term:

> Most companies with an inside-out perspective become attached to what they produce and sell and to their own organizations. In contrast, the outside-in perspective starts with the marketplace and delves deeply into the problems and questions customers are facing in their lives. It then looks for creative ways to

combine its own capabilities with those of its suppliers and partners to address some of those problems. The goal is to bring value to customers in ways that are beneficial for them while also creating additional value for the company itself.

The belief that to reliably achieve "value for the company itself," you have to first discover the best ways to "bring value to customers" is exactly what Iger missed in his claim of "significant transformation." Iger's words reflect an inside-out perspective when he should have offered more grounded reassurance. He could have made it clear that he and Disney's other executives were acutely sensitive to the motives or other "outside" factors that moved so many subscribers to abandon the service in the first place.

As seasoned business professionals, we are more than familiar with the terms B2B, B2C, and DTC. But what if we flipped the narrative? Let's call it "C2B—Customers go to Businesses," or maintaining "B2C," let's try"Buyers to Companies." This shift to the consumer perspective is part of the rationale behind the increase in direct-to-consumer (DTC) business models. Research predicts that the DTC online market will reach $213 billion by the end of 2023, and more than 60% believe DTC provides the best personalized and most engaging digital experiences compared to traditional brands.

Recently I participated in a major film and television studio's workshop, which included the company's C-suite and several business unit leaders. The objective was to identify opportunities where they could transform their technology and improve business operations. During the event's first hour, we listened to different perspectives on each leader's problems and discussed the studio's goals.

We struggled to move past this topic once it became clear that everyone had different interpretations of the studio's primary goals, and individual leaders had biased opinions about their departments' needs. The only common theme was the desire to modernize their technology in order to make their work lives easier.

The debate over the studio's central goals lasted 45 minutes, with one executive after another chiming in. I sat quietly for most of the conversation until I sensed an opportunity to challenge the room and asked, "Can anyone tell me how your individual or studio goals relate to the customer?" There was a long silence until the CIO replied, "Our customers are other businesses like Netflix or movie theaters. We're B2B and don't target the consumer directly." I responded, "Well, that is part of the problem."

The ensuing discussion included several rejections of my premise, but I pressed the issue further, explaining that it was vital that the studio understand what the consumer wanted. Recognizing that their direct customers are other businesses (B2B) focusing on contracts, sales, and content distribution, does not discount that those businesses have customers (B2C) that ultimately connect back to the studio. Combining the two creates a B2B2C model, but the outside-in perspective reverses it to C2B2B.

The conversation went on for another 45 minutes, and although most of the people in the room were not on the same page, the CFO, the top industry veteran in the room, looked at me and said, "You're right. The money flows from our fans to our customers to us. If our fans dislike our content, our customers drop the contract, and we lose business."

As we reflected on implementing modern technologies to make their work lives easier, I suggested they leverage that same technology for a deeper understanding of the fan experience in order to build a relationship and grow their business. The VP of Product Development challenged me, "Remember we are B2B, and we don't look at the customer journey. To us, the fan experience starts and ends when they see one of our movies."

I asked the room to take a moment to reflect on the movie-going experience and explained, "The customer journey starts before we get to the theater and continues after we leave." From the outside-in perspective, the complete journey is something like this:

We decide to see a movie, select a film and buy our tickets online or at the box office. At some theaters, we might even reserve our seats. If you're like me, you calculate time for snacks, traffic, trailers, and credits, and leave the house accordingly. We arrive at the theater, and the usher scans our tickets. We may use the restroom and grab additional snacks, popcorn, hot dogs, or soda. We find our way to the auditorium, take our seats, watch trailers, and, hopefully, enjoy the movie. But our experience doesn't end when the movie does. We leave the theater, tell our friends and family about the movie, post our opinions on social media, and plan to see it again if it's worthy of a second viewing.

The group around the table at the workshop displayed various expressions of agreement and sentimental recollection. Within a short time, the participants could visualize the customer-first perspective and connect it to their business goals. I suggested relating snack purchases to advertising campaigns, historical theater seating preferences to promotional timing of ticket sales, and social media conversations to fan loyalty. The CIO turned and looked directly at me and asked, "How do we do that?"

The real opportunity lies in aligning your mindset with an outside-in perspective. The same technology the studio wanted to adopt to improve their business processes is better positioned to gain insights on their customers' customers first.

Adopting an outside-in perspective means considering each business decision or proposed action by looking outside the company at your potential and existing customers and making them your primary concern, prioritizing their experience and satisfaction. This approach seeks to understand the customer's journey and, more challenging, their beliefs from their point of view. This perspective is shaped by customer engagement with the marketplace, your brand, and the community of loyalists they are a part of. The challenge here is understanding who your customers are as need, preference-motivated, goal-driven consumers. Doing so is imperative for success, as Thomas von Ahn reminds us:

> If you don't understand who your customers are—not just that but what they do in their spare time, how much money they make, what issues they have, where they shop, what devices they use, and which way they like to hang their toilet paper roll (OK, maybe not that, but you get our point)—then it's almost impossible to succeed.

Adopting an outside-in perspective requires businesses to rethink how they approach their operations and services. Instead of starting with *what the company can offer*, businesses must begin by asking *what the customer wants* and how the company can deliver a solution that meets those desires in the best possible way. This requires companies to have a deep understanding of their target audience, their motivations, and their preferences. Like the studio workshop, you have to change your mindset about where the focus is.

Pixar's Chief Creative Officer, Pete Docter, said in an interview that the *Lightyear* movie failed because "the film simply asked too much of the audience." *Lightyear* had a production budget of $200M and made $226M worldwide, compared to *Toy Story 4*'s $1B. He continued acknowledging, "We've done a lot of soul-searching about that because we all love the movie, and we love the character and the premise." Notice the consistent use of "we" in his explanation. What comes next, however, is gold. He continued, "I think probably what we've ended on regarding what went wrong [with the *Lightyear* movie] is that we asked too much of the audience. When they hear Buzz, they're like, 'Great, where's Mr. Potato Head, Woody, and Rex?' And then we drop them into this science-fiction film that they're like, 'What?'" Notice, Docter's response has now shifted to "they" representing the audience. But it's his final comment that summarizes the issue. He added, "I think there was a disconnect between what people wanted/expected and what we gave them."

One of the key benefits of an outside-in perspective is that it helps companies create more value for their customers. In the long run, it will also increase the company's value. As von Ahn states, "An outside-in strategy is unique because it uses customer trends to guide what products and services are offered." Adopters of this strategy constantly strive to stay attuned to market dynamics. Companies that adopt an outside-in perspective consistently seek ways to stimulate demand for their products or services by aligning with their customers' needs and behaviors. Their approach, as explored in Chapter 2, enables them to think in tandem with, or even one step ahead of, their customers, ensuring they consistently deliver high value. Furthermore, this mindset can be adopted by all of us. By researching our customers and the broader market, we can become conscious of trends and adjust our strategies to anticipate future customer needs.

This leads to higher customer satisfaction and loyalty with increased brand reputation and trust.

Another advantage of the outside-in perspective is that it encourages innovation. By consistently placing the customer at the center of business decisions, companies are forced to think creatively about better serving their customers. This perspective can lead to new and unique products, services, and business models that set the company apart from its competitors. Along these lines, von Ahn suggests that companies with an outside-in strategy are more prone to consider:

How they can solve today's problems in a new way
What opportunities are presenting themselves
What needs aren't being met in the market

This problem-identifying, solution-oriented approach fosters creativity and innovation.

A Better Way to Hail a Cab

An excellent example of a transformative application of the outside-in approach is the emergence of Uber. As this Cascade strategy study documents, Uber was conceived with a distinctively outside-in mission focused on the needs of their customers: "*to make transportation as easy to access as running water* and . . . to do it differently, without owning its vehicle fleet like your regular taxi company."

Uber's journey began with a classic origin story in which its two founders, Travis Kalanick and Garrett Camp, encountered a familiar problem while attending a technology conference: they could not successfully hail a cab. Realizing that countless millions of other urban dwellers had experienced this same predicament, the duo asked themselves, "What if you could request a ride from your phone?" This idea led to the eventual creation of Uber, one of the most disruptive forces in transportation around the globe. As the strategy study concludes, Uber's success has primarily revolved around its ability to identify and meet the needs of two key customer groups: riders and drivers. Uber's fundamental concept is to bridge the gap between passengers needing a ride *and* drivers ready to provide one. Passengers generate the "demand," drivers provide the "supply," and Uber facilitates this effortless interaction. Uber relies on two primary user groups and must present compelling value propositions to drivers and passengers to ensure sufficient user engagement for the platform to operate effectively.

At nearly every turn, Uber made critical strategic decisions in direct response to the needs and preferences of its growing customer base. The same strategy study points out that Uber's early customers became such passionate supporters of the company because Uber was advocating for them. That is, Uber not only recognized riders' frustration with the business-as-usual attitude of the transportation industry in their cities but actually did something about it and actively "fought against old regulations" on behalf of the customers. Recognizing and nurturing these customers from the get-go was an astute decision. By placing customer convenience and service at the forefront, Uber embraced the "disruptor" role and incorporated it into the company's identity and branding. They became part of a broader socioeconomic movement focused on revolutionizing traditional industries to serve consumer needs better.

Even after problems came to light regarding toxic elements within Uber's corporate culture, resulting in Kanalick's resignation as CEO, Uber's rebranding efforts featured a distinctively outside-in perspective, as "[l]istening, observing, and learning became the foundations of Uber's cultural overhaul." In reference to Uber's drivers (now called "driver-partners"), the new CEO Khosrowshahi observed that "[w]e talk to them, we have a dialogue with them, and we build with them. That kind of connectivity with our driver-partners, I think, creates a win-win, and it creates mutual respect."

Implementing an Outside-In Approach

Like most aspects of an effective business strategy, an outside-in perspective must be intentionally and methodically cultivated within a company's culture and decision-making processes over time. Typically, this process will involve a complete company mindset shift because, as discussed in Chapter 1, business and technology leaders often believe they already know what is best for customers or have biased opinions about the technology they aim to develop.

This implementation of an outside-in perspective within a company can be broken down into 6 main elements:

1. Customer research and analysis. The outside-in perspective and the customer transformation process generally begin with thoroughly understanding your target customers. Uber's success was due in significant part to its willingness to profoundly understand its customer base, including both drivers and riders. This requires humility, customer empathy , and the ability to determine significant insights about your customers. In terms of method, such understanding can, in most cases, be developed only when a company is willing to invest in ongoing customer research and the regular collection and analysis of feedback. This may involve surveys, focus groups, and numerous other forms of customer engagement. The results of these efforts can then be used to inform product development, improve customer service, and optimize marketing strategies.

The good news is today's technology provides unprecedented opportunities for businesses to connect with customers, making understanding what drives them more feasible than ever. The emergence of novel marketing techniques has enhanced opportunities to engage with our customers and deliver more value. Among these are omnichannel marketing, personalization, and social Customer Relationship Management (CRM), to list a few. Utilizing these resources optimizes our ability to understand and connect with our customers, enabling us to tailor our marketing strategies effectively.

The opportunities for this direct connection with customers are perhaps most abundant through social media. As one web development firm observes, "[w]e live in an age where many people feel more comfortable sharing their honest answers to questions from behind a screen." Social media provides numerous ways for companies to elicit such forthright feedback. While integrating forms into social media and offering brief surveys for visitors are straightforward methods of gaining feedback, direct engagement is the most effective way to obtain authentic, valuable insights. Conduct personal interviews with your customers to understand their expectations and gauge how your offerings stand against them. Encourage customers to post testimonials on your platforms. Whenever an issue arises, resolve it publicly within the platform to benefit your entire follower base. We'll discuss these types of communities in more detail in Chapter 8.

It would go beyond the scope of this book to explore the many ways in which technology—including social media—can be used to understand your target customers. Let me emphasize that the understanding needed for a practical outside-in approach to customer transformation must be more than superficial. This will likely require new and creative methods of data collection, as Harvard Business School's Ranjay Gulati notes:

> Gaining real insights into customers' needs demands more of companies than those arising from typical market research. The questions you ask of customers must be more profound and open-ended, with an intent not only to discover how your customers engage with your products or services but also to understand some of the broader parameters of the constraints they are facing in their own lives.

It's important to recognize that merely soliciting feedback from your customers may only sometimes provide you with comprehensive insights. As the famous quote attributed to Henry Ford illustrates, "Had I asked my customers what they desired, their response would likely have been *a faster horse*." In other words, it's not enough to gather data—even lots of data—about your customers. The data must be thoughtfully interpreted within the current marketplace context, leading to a thorough understanding of the motives and goals that drive your customers—including identifying needs before your customers even ask for them.

2. Customer-centric decision-making. As you better understand your customers, it's also vital to directly educate your employees and executives to bring about the shift in corporate culture needed to establish an outside-in perspective. Preparing for and achieving this shift will involve prioritizing customer satisfaction and experience in all business decisions. This requires encouraging employees at all levels of the organization to think from the customer's perspective (i.e., practice empathy toward the customer) and prioritize customer needs when making decisions.

3. Align business operations with customer needs. This third element of the implementation process is essentially where the rubber meets the road: Your company will need to undertake a systematic reevaluation of your products, services, and processes to ensure that each is aligned with customer needs and expectations, then make changes as necessary to optimize the customer experience.

4. Measure and track customer satisfaction. Although this aspect of implementing an outside-in perspective could be under customer research, it's worth mentioning separately to emphasize the fact that understanding your customers is an ongoing, never-ending process. This understanding doesn't end after you've finished your initial research, effected a shift in the company's mindset, and made changes to align your products and services with the customer's needs. Not only might your initial understanding of your target customers require adjustments, but your customers will also inevitably change over time, which means their satisfaction with your product or service may fluctuate. For this reason, it's crucial to regularly measure and track customer satisfaction through metrics such as Net Promoter Score (NPS) and routine customer feedback. Use this data to continue identifying improvement areas and making necessary changes.

5. Empower employees to act. The responsiveness to customer feedback called for above will make a real difference only if the relevant employees are empowered to act on what they're learning from customers. Therefore, it's crucial to provide employees with the resources and training necessary to respond to customer feedback and make appropriate changes to improve the customer experience.

6. Continuously improve the customer experience. Every element of the implementation process mentioned above has pointed to the goal of remaining focused on delivering a great customer experience and constantly seeking new ways to meet customers' needs better.

SIX

EXPERIENCES IN THE MOMENT

"Poets are like baseball pitchers. Both have their moments. The intervals are the tough things."

Robert Frost

The Devils Tower Bacon Cheeseburger Challenge

During the summer of 2022, my 15-year-old son and I embarked on a road trip from Southern California to Mount Rushmore, then to Montana, through Yellowstone, and eventually to Iowa, where we fought the foamy rapids of a whitewater river. In total, we drove 3,400 miles over the course of seven days—an unforgettable journey.

At one point, we stopped at Wyoming's Devils Tower, a national monument most recognized as the centerpiece of Steven Spielberg's 1977 movie *Close Encounters of the Third Kind*. Nearby in Hulett, we grabbed lunch at the famed Red Rock Café, where my son eagerly accepted the Devils Tower Bacon Cheeseburger Challenge: consuming three 6-ounce burgers with cheese, bacon, and the works. I was content to watch, as were several tables of other smiling onlookers, as in surprisingly short order, my teenage son devoured the entire plate—burgers, fries, and all.

Sitting at the table next to us was a friendly couple who cheered my son on and continued chatting after the eating experience. It turned out, Bob and Louise were a retired couple traveling through the area on their Harleys. Several minutes into the conversation, we discovered a remarkable fact: Bob and Louise's home was less than 2 miles from our own back in Southern California! It was an incredible experience to meet neighbors 1,300 miles from home whom I might never have met back in our neighborhood.

That experience in the diner was just one in a series of moments that my son and I will always remember as our Summer of 2022 Road Trip. But there's something particularly interesting about the diner encounter. While most of us meet or interact with new people nearly every day, the vast majority are unremarkable. What made the interaction with Bob and Louise a highlight of the journey was how it *connected* that Road Trip journey to our lives in Southern California. In a sense, we're all living out multiple life "journeys" every day through the various long-term goals, projects, and relationships we pursue. This diner experience in Wyoming connected my son's and my "Summer of 2022 Road Trip" journey to our bundle of California-based "journeys" in a unique way when we discovered Bob and Louise were neighbors. This connection—in addition to the crazy odds that we would happen to meet up with neighbors so far from home—made the diner experience even more memorable.

This chance encounter also opened up an array of possibilities. For example, given how well our conversation was progressing, it wouldn't have been surprising if, at some point in the conversation, either Bob or Louise had pulled out their cell phone and started showing us photos of their grandchildren or narrating funny stories about their Frenchie back home. Or they might have offered to step out into the parking lot with us and show us their Harleys, or perhaps we might have enjoyed their company so much that we would've asked if they wanted to accompany us that afternoon to visit Devils Tower. Or we might have exchanged numbers and made plans to connect once back in California—a possibility that might have spun off into an entirely new set of experiences with this couple. Whether we did any of this is beside the point: these and many other hypotheticals became possible through the power of that simple connection made in the diner.

What does this story have to do with customer transformation? Consider this: Our customers are also in the midst of multiple journeys, some of which we as businesses help curate and are often referred to, unsurprisingly, as "customer journeys." The problem is many companies tend to be so focused on developing these journeys that they lose sight of the many "connections" their customers are experiencing through intersections with *other* trips being curated by other businesses and a multitude of other agents within the broader society and culture. The result is a kind of myopic vision in which companies miss out on opportunities to significantly enrich the customers' experience with their brand and thereby deepen customers' sense of brand loyalty.

But this doesn't have to be. Just as my son's and my connection with a retired biker couple in a Wyoming diner became a uniquely memorable moment with the potential to enrich several otherwise disconnected journeys, your business can anticipate and facilitate connecting moments for your customers in ways that maximize the benefits for them and your business. This chapter is about identifying and developing such connected moments and, in so doing, making productive, satisfying connections between our customers' larger journeys.

Fundamental Concepts

But first, we need to get some basic concepts and definitions under our belts. Let's start with three concepts: moments, experiences, and journeys.

Moments

A "moment" within the broad context of customer experience is *a specific interaction or touchpoint that a customer has with a brand, product, or service* (recall the "moments of action" from Chapter 3, with the use of technology serving as "people interfaces" to enrich our customers' experiences in authentic ways). These moments can significantly impact the quality of a customer's overall experience with and perception of your business and brand.

In order to define moments for customers, it is helpful to categorize them into four general types based on their purpose and effect on customer experience.

Informative moments

In these types of moments, the customer becomes more familiar with your brand, product, or service. Such moments can occur through various channels such as advertisements, social media, blog posts, or word-of-mouth. Informative moments aim to provide customers with valuable information that helps them better understand your brand's offerings and the benefits they may bring.

Engaging moments

These moments involve direct interactions between customers and your brand, whether online or offline. Engaging moments can include when a customer visits your website, uses your mobile app, interacts with a customer support team member, or attends a brand event. These moments are crucial for building customer connections and fostering a positive relationship.

Decision moments
These are critical moments when customers evaluate their options and decide whether or not to purchase a product or service. Decision moments may involve the customer comparing products or services, reading reviews, or seeking recommendations from friends or family. Your business should strive to make the decision-making process as smooth and straightforward as possible for the customer.

Part of decision-making moments include the amount of effort required from the customer during their experience. Content gates, lengthy forms, and multiple steps often cause frustration that generate moments where customers may abandon the process.

Delightful moments
These moments are unexpected, positive experiences that exceed your customers' expectations and create lasting impressions. Delightful moments can include personalized offers, discounts, surprise gifts, or exceptional customer service and may significantly contribute to customer satisfaction, loyalty, and advocacy.

By understanding and categorizing these different types of moments, your business can begin to design and optimize positive, memorable interactions with your customers. However, to engineer a memorable customer moment, it's important to consider that moment's wider context—which is where many businesses fail. Many external factors can affect how a particular customer moment is experienced, so let's examine some of the most important of these external factors.

Economic factors

Conditions such as inflation, unemployment rates, and consumer spending can affect customers' purchasing decisions and experiences with your brand. During an economic downturn, customers may become more price-sensitive, impacting their perception of your brand's value proposition.

Societal factors

Trends, cultural shifts, and societal changes can also influence customer moments. The rise of social media has changed how customers discover and engage with brands and how they share those experiences with others. Additionally, increasing awareness of environmental and social issues may impact customers' preferences and expectations, causing them to favor brands that align with their values.

Technological factors

Advancements can lead to new customer moments and touchpoints. The widespread adoption of smartphones has created opportunities for brands to engage with customers through mobile apps, location-based services, and augmented reality experiences. Likewise, emerging technologies like AI and IoT can lead to new types of customer interactions and alter customer expectations of their interactions with your business.

Competitive factors

The actions and strategies of your competitors can also influence your customers' moments. If a competitor launches a new product or service offering a unique value proposition, some existing customers may re-evaluate their loyalty toward your brand. Additionally, competitors' aggressive marketing campaigns or pricing strategies can impact customers' perception of your brand's offerings and value.

Legal and regulatory factors
Laws and regulations can affect customers' interactions with your business, particularly in highly regulated industries like finance, healthcare, and telecommunications. Compliance with these regulations (such as collecting information for security or privacy purposes) can impact the tone and content of interactions, potentially creating friction in specific customer interactions.

Weather and climate
These types of conditions can momentarily impact customer behavior and experiences. Heavy rain or extreme temperatures may discourage customers from visiting your physical stores or attending outdoor events. At the same time, temperate conditions can boost foot traffic or increase demand for specific products or services.

As policy makers strive to implement strategies and initiatives to alleviate the impacts of climate change, the business sector too is acknowledging their part in addressing environmental and sustainability challenges— a responsibility that has become increasingly significant with customers.

Time constraints
A customer's available time can significantly impact interactions with your brand. When making a purchase decision, a rushed customer may prioritize speed and convenience over other factors, such as price or product selection.

Geographical location
A customer's spot on the map can influence the nature and extent of their interactions with your brand. Customers in rural areas might have limited access to brick-and-mortar stores or face longer delivery times for online purchases, affecting their overall experience.

Life events
Milestones such as a job change, moving to a new city, or the birth of a child can influence a customer's needs and preferences, potentially impacting their relationship with your brand.

Emotional state
A customer's emotions can affect how they interact with your business. A customer experiencing stress or sadness may be more sensitive to negative experiences or more appreciative of joyous moments.

Social context
The voices of friends, family, or colleagues can influence a customer's decision-making process and interactions with your brand. A customer may be more likely to try one of your products or services if a friend has recommended it, or they may be influenced by the opinions and preferences of their peer group.

One important characteristic of the external factors listed above is, by definition, they typically originate entirely apart from customers' interactions with your brand and are, therefore, largely out of your company's control. For this reason, these factors are simply off the radar for many companies. While external factors may be out of your company's control, you should not ignore how such factors influence potentially significant customer moments.

Experiences
"Customer experience" is a widely used term that refers to *a customer's perceptions, emotions, and reactions when interacting with a brand, product, or service.* The concept encompasses the customer's positive or negative perception of the brand, the perceived usability and functionality of the company's products and services, and the emotional impact that interactions with the brand and the use of its

products and services have on the customer. In the words of Forbes Technology Council's Cameron Weeks, customer experience is "about how customers feel when interacting with a brand." As such, customer experience (CX) is closely related to the concept of user experience (UX), the latter referring more specifically to a tech user's perceptions of how positive, intuitive, and seamless the technologies are that enable the user to interact with a brand.

The phrase "customer experience" can reference a customer's feeling about a particular interaction with your brand (e.g., at a particular touchpoint) as well as the broader synthesis or aggregate of their perceptions of your brand based on the totality of significant moments they've experienced with your company up to the present point. Customer experience thus encompasses all aspects of the customer's interactions with your company, including all touchpoints across various channels such as websites, mobile apps, social media, or in-person interactions. The customer experience is influenced by factors like the quality of your products or services, the level of customer support experienced, the ease of use of your products, and the customer's overall satisfaction with your brand. The main goal of customer experience management is to provide seamless and enjoyable encounters with customers, which can ultimately lead to brand loyalty and positive word-of-mouth.

Journeys

The "customer journey" is another commonly used term and refers to a (typically visual) representation of the complete path that a customer follows in their interactions with your brand—particularly in relation to making a purchase—from the point of initial brand or product awareness through the purchase decision and on to post-purchase experiences. Notice that whereas the customer experience refers to the customer's perceptions, the customer journey is characterized in terms of the typical customer's observable behaviors or actions, or in Weeks' words, "what customers do at each stage of their life cycle: before, during and after they buy from a brand" (emphasis added). The customer journey maps across all touchpoints and stages a customer experiences such as awareness, consideration, evaluation, decision, purchase, and post-purchase support. Thus, it is a helpful tool for businesses to map out and understand customers' interactions and experiences with their brands. By analyzing the customer journey, businesses can identify areas of improvement, optimize touchpoints, and design targeted marketing strategies to enhance the overall customer experience.

Imagine your business has fashioned a particular customer journey to include a path along points [A→B→C→D→E] that, from the company's perspective, might refer to your customers' interactions with the brand at different touchpoints:

Email→Website→Catalog→Purchase→Completed

The development of this journey will have included numerous design decisions on your part to encourage a particular customer experience of the email newsletter, supplemented by yet another intended customer experience for the website, and so on. In most cases, a company is only looking at this journey from the perspective of their "typical" or "target" customers and of how the business hopes the customer will respond at each point, which means, the individual experiences of actual customers within that process may vary. So, one customer's experience through the touchpoints of this journey might be:

tantalizing→unappealing→difficult→ straightforward →secure→reassuring

But another customer's experiences within that same journey may be different: they may find the website sleek and professional but the checkout process frustrating.

Take a Little Trip

With that in mind, let's return to the idea of "connections" so we can begin to understand how customer transformation comes into all this. When commenting on the great Summer of 2022 Road Trip, not only is each journey that your customers' experience made up of many separate significant moments—each potentially curated by your company—but your customers also are embarked on multiple other journeys at the same time, some of which have nothing directly to do with the products or services your company is promoting. Furthermore, these different journeys can intersect at particular moments in time, such that a customer's experience from one journey can influence their experience of another during that intersection. The risk is that your company may be so focused on the one or more journeys you're actively cultivating that it overlooks the significance of these connections or intersecting moments customers have with journeys external to your own.

Consider a typical business travel scenario. The traveler will likely have to interact with multiple agencies through various technologies and will, therefore, actually embark on multiple customer "journeys." A typical (oversimplified) sequence of these overlapping journeys might include:

Airline→Uber→Hotel→Uber→Restaurant→ Calendar→Meeting→Uber→Hotel→Uber→Airline

Within the airline customer journey, the traveler may

pass through several significant touchpoints, including the airline's website, mobile app, check-in counter, etc. There will be another journey with different moments (although some may be shaped similarly) with the taxi company and another with the hotel, with the restaurant, and so forth. Typically, these separate journeys are siloed, as the different agencies or companies involved are focused on each of their journeys, despite the apparent fact that they share many customers.

Now, imagine a world in which the technology used to facilitate each of these separate journeys understands how they overlap and intersect at significant moments. Imagine, further, that these technologies are responsive to what came before—in these *other* journeys—and proactive about what comes next. Suppose a traveler has a terrible airline experience (i.e., a flight delay) and that—instead of the traveler having to scramble to adjust the Uber pick-up time, their hotel check-in, their dinner reservations, etc.—the technologies for the customer's Uber, hotel, restaurant, etc., will take the new timetable into account, automatically making whatever adjustments are necessary to provide a seamless experience. Imagine that the Uber's arrival at the airport will be spot-on, there will be an available bellhop ready to assist the traveler at the hotel check-in, the coffee awaiting the traveler in their room will be freshly brewed, the thermostat will have engaged precisely in time to make the room match the traveler's preferred room temperature, the table reservations at the restaurant will be pushed back accordingly, and so on. The possibilities are almost endless when you begin to think of the many potentially significant moments that might be curated for the traveler across these intersecting journeys to produce an overall more satisfying customer experience.

Why isn't something like this done more often? Until recently, it was relatively difficult, if not impossible, to pull off this sort of seamless customer experience, as the technology hadn't arrived. But with advances in AI and data management, that is increasingly not the case. There's also the issue of competition between different entities, but in many cases, the potential for collaboration and partnership would appear to outstrip the risks of competition. We could do all of this and more today, but due to apathy or for other reasons, businesses have not wanted to, at least not yet. One of the goals of this book is to break down the silos between the various customer journeys and enable businesses to deliver more satisfying customer experiences, boosting brand loyalty and increasing profitability along the way.

Baseball's Fan Experience

Major League Baseball (MLB) has undertaken the inspiring endeavor of making moments of the game feel engaging and alive. They are innovating the very experience of the game while simultaneously knitting together a 'connected' baseball community that transcends geographical confines.

MLB has boldly ventured into the digital realm with the Automated Ball-Strike system (ABS). The goal was to validate whether a system could make and consistently correct calls to the umpire quickly enough for the umpire to make the call on the field without introducing a delay. The results of the work in 2019 confirmed that the system could work with sufficient consistency, reliability, and responsiveness.

Augmented Reality (AR) has also become a tool to engage fans who might be physically distant from the stadium. It has brought the field into living rooms, where fans can visualize player statistics in 3D, compete in a virtual Home Run Derby, and immerse themselves in friendly trivia games during pitching changes. The intervals between active plays come with dynamic, engaging experiences in the comfort of one's home.

MLB is also exploring the captivating world of digital collectibles with Non-Fungible Tokens (NFTs), creating a new avenue for fans to connect with the game's past, its iconic moments, and their beloved players. Fans can now digitally possess a piece of baseball history, a perfect blend of nostalgia and innovation.

Their technological advancements are revolutionizing the fan experience by turning the 'tough intervals' into moments brimming with engagement and excitement. MLB has connected to the digital age creating engaging fan experiences in every moment of the game.

ACTION PLAN

STAGE THREE: JOURNEYS

Introduction
The third stage is dedicated to developing a deep understanding of customer journeys and experiences. The goal is to unlock the power of the outside-in perspective to enhance these journeys, delivering personalized experiences that resonate at the moment. It's about customer journey mapping and understanding how journeys intersect, connecting your customer before and after they directly interact with your business. Trust in customer transformation means recognizing and fostering these interconnected journeys.

Day 0: Reflection and Goal
- **Reflection:** Reflect on your current understanding of your customers' journeys within your business. Consider the experiences you've crafted for them and how personalized these are. Think about the intersections between different customer journeys and how these intersections contribute to your overall customer experience.

- **Goal:** Your primary goal for this stage is to understand your customer journeys deeply and strive for an outside-in perspective. We aspire to recognize and enhance the moments where customer journeys intersect, creating a continuous and seamless experience that transcends singular interactions. As we explore customer journeys, we aim to initiate a mindset centered around personalization, interconnectedness, and the power of understanding your customers' paths. This mindset will lay the groundwork for improving satisfaction, engagement, and customer lifetime value.

Day 1: Content Walkthrough
We will delve into the concept of customer journeys, the power of the outside-in perspective, and the role of customer transformation in building these journeys. We will also discuss how to provide personalized experiences that make a difference.

Day 1: Workshop Questions
1. How can an outside-in perspective transform our current customer journeys?
2. What role does personalization play in enhancing customer experience?
3. Can we give an example of interconnected customer journeys between us and other businesses?
4. How can understanding "a collection of journeys" lead to better customer transformation?
5. How might moments of action influence loyalty and engagement four our customers?

Day 2-7: Homework and Next Steps

Over the following days, reflect on your customers' journeys, identifying opportunities for personalization and recognizing intersections. Consider how an outside-in perspective could enhance these journeys and transform the customer experience.

Day 8: Touchpoint and Next Steps
1. Share your reflections on the current customer journeys.
2. Discuss ideas for enhancing personalization and improving the customer experience.
3. Identify opportunities to better interconnect different customer journeys with other companies.
4. Develop an action plan based on these findings.
5. Schedule the next meeting to review and refine this action plan.

Day 9 and Beyond: Action Plan

30-day plan: Begin enhancing your customer journeys by incorporating personalization and an outside-in perspective. Considering all touchpoints and stages, map out the customer journey from initial brand awareness to post-purchase support.

60-day plan: Implement changes based on your new understanding of the customer journeys and measure the impact on the customer experience.

90-day plan: Evaluate the effectiveness of the changes and the impact on customer relationships. Refine your approach based on the results and feedback, and prepare for the next training phase.

STAGE FOUR: ECOSYSTEM

Participate in Community Evolution

SEVEN

CHANNEL-LESS INNOVATION

*"The only way to discover the limits of the possible
is to go beyond them into the impossible."*

Arthur C. Clarke

Apple's App-eal

Remember when a 2009 Apple ad launched the "there's an app for that" slogan that grew to a phenomenon? The phrase caught on like wildfire and remained in our collective consciousness for years afterward, spawning everything from a Tedx talk to a "Sesame Street" song. One reason the phrase was so remarkably catchy is that it shone the spotlight on a new, fascinatingly decentralized approach to innovation. Any entrepreneur with a clever idea for enabling the iPhone (and soon other app-enabled devices) to help us explore, document, or otherwise manage some facet of our everyday lives could market that tech through an app store and potentially generate billions in new business. It was a huge step toward moving innovation from something that occurs primarily in corporate labs and meeting rooms to something that can percolate among the masses, as it were. The explosion of apps over the past decade-plus is just one example of a wider phenomenon in the modern world, one in which the centers of information, influence, and agency have shifted from relatively few (often tightly controlled) hubs to platforms that are more broadly dispersed. Think of wikis or of DeFi (Decentralized Finance), or even the concept of a news item or other social media post "going viral." In the latter case, the idea or content in question quickly moves beyond the reach of its initial source and, through the agency of numerous individuals, jumps from one channel to the next (i.e., social media platforms, types of devices, etc.) in ways that transcend any of the particular channels involved.

Changing the Channel

In Chapter 5 we discussed today's customers' expectation of a seamless experience across various touchpoints and channels when interacting with your brand and, indeed, between your brand and the other agents within their customer ecosystems (i.e., other brands, distributors, customers, etc.) Recognition of this fact and the willingness to adopt a holistic perspective toward customer interactions, I argued, is crucial for fostering a successful customer transformation strategy.

This perspective is closely related to a popular perspective known as the "omnichannel" approach to the customer experience, along with a more recent successor, the "channel-less" approach. Omnichannel refers to the centralization of data "across all mediums" so that customers can "therefore expect a consistent experience across all touchpoints and interactions with an organization." The channel-less approach takes this one step further by seeking to remove the focus on channels altogether: "Channel-less communication takes place when customer communication across multiple channels is linked into a single smooth conversation. The channels themselves become irrelevant—creating a cohesive customer experience." As Shalaka Nalawade observes, "It's the marketer who thinks in terms of channels, not the customer." This perspective is not as strange as it may seem at first glance argues Claire Lucas, Head of Product Marketing at Odigo: "We are all channel-less in our private lives. A son emails his mother birthday greetings; she replies by phone, and when she asks for the latest photos of his family, he follows up via social media. The channels are not important; the development of the conversation is what matters." Again, this notion of "channel-less" customer interaction goes hand-in-hand with what we discussed in Chapter 5 about engaging customers with a full awareness of the ecosystems they seamlessly inhabit and of the journeys, experiences, and moments they have within those ecosystems.

In this chapter I propose that something similar applies to your company's innovative process. Let's call it *channel-less innovation* and define it in this way: *Channel-less innovation means innovating without being limited to particular channels or particular sources of ideation*. Such innovation is inspired and directed not by channel-specific technology but by the broader "conversations" in which customers are engaged within their ecosystems. Channel-less innovation thus transcends any particular channel and, therefore, opens the door not just to external business partners making contributions that will benefit your company's innovation process but also to the customers who play a vital role in that process. This moves you beyond the traditional sources of input for inspiration (e.g., internal teams, employee suggestion boxes, etc.) and allows you to leverage direct customer involvement in innovation—with the potential of developing an army of influencers generating and developing new ideas with tools provided by your business. Partnering with the customers themselves allows the business to scale innovation outside company walls, which can foster deeper customer connections, driving loyalty and enhancing customer satisfaction.

Customer Ecosystems

An additional concept that is receiving increased attention and is directly relevant to the smart collaborations described above: *customer ecosystems*. The customer ecosystem is a broader perspective on the customer experience that emphasizes the interconnectedness of various brands, products, services, and touchpoints a customer encounters throughout their many journeys. Acknowledging these ecosystems means recognizing that your customers interact with other brands and experiences beyond your own specific offerings and that your customers have expectations, preferences, and habits shaped by their experiences with these other brands and agents—even those in unrelated industries.

Graham Hill, Associate Director of Optima Partners, explains it this way:

> The customer sits at the centre of their own ecosystem of providers . . . Most companies only look at their [own] interactions with the customer, . . . [and t]hey use this understanding to improve the interactions in the pathway from their own perspective. They think that is how the customer sees things. But that isn't how the customer sees things at all. In addition to the interactions with the company, the customer interacts with many other actors to get the resources they need

to get their jobs done. The customer looks at all the interactions in their journey and the different actors involved; at the ecosystem. This is a very different perspective.

Unfortunately, most businesses today focus solely on the events within their own "walled garden" and offer a curated experience only within that controlled environment. For example, when you shop online with your favorite retailer, they may be concerned with the experience they directly provide but rarely consider what happens before or after you leave their online store environment. You may see some connections beyond the store, if your online purchase is handed off to a separate shipping and delivery company so that the experience for the customer continues through a third party. But typically, in such cases, the seller and the shipper have minimal influence on each other's journeys fashioned for the customer.

Such siloing is unrealistic. As Mark Taylor, Sr. VP and Global Practice Lead for Cognizant Digital Experience, reminds us: "Companies don't fully own their experiences. In the complex experience ecosystem, partners and customers deliver parts of the experience, and this reality requires companies to think about how they can enable everyone involved in experience delivery." Adopting this holistic view of the customer's ecosystem allows your business to identify opportunities for collaboration, partnership, or co-branding to create more comprehensive and seamless experiences for your customers. It also enables your company to learn from best practices and innovations employed by other brands, which can help improve your customer experience.

Customer 720

Awareness of customer ecosystems is closely related to what has been coined the "Customer 720" view, which is itself an extension of another concept that has been popular for the last decade or so, the "Customer 360" view. Rahim Hajee, Chief Technology Strategy Officer at Adastra, explains that the goal of Customer 360 is to gain a better understanding of a company's customers by keeping track of all relevant internal data, including "how many your organization has, which accounts belong to your customers, what products they have, how to contact them, preferred channels, who their internal representatives are, and contractual information."

Going beyond this, a Customer 720 perspective takes into account external customer data as well and "maps your customers to their social media profiles (e.g., Do they have a LinkedIn profile? How active are they? Are they tweeting about your brand?), loyalty information, channel preferences, and marketing campaigns." Hajee explains how this information can then be put to use:

> [L]everaging deep AI and ML-driven segmentation to drive analytical capabilities, . . . we can . . . assess each interaction and its impact on key KPIs, revealing the next best actions to take to enhance customer loyalty and brand affinity and get to a deeper level of personalization. For example, if they had a negative experience, perhaps there are further discounts that can be automatically applied to their current transaction, or if they are repeat customers with a high lifetime value, perhaps advanced notifications of new merchandise or additional bundling savings are target campaigns that can resonate with the customer.

The benefits of such a holistic 720 view include, according to Hajee:

- "Personalized offers for customers (e.g., dynamic pricing, subscribe-and-save models, etc.)
- Understanding customer friction points to improve their journeys
- Cost optimization of customer retention/acquisition
- Reduced $ spent on research and development for new products."

The data that a 720 view seeks to manage, assess, and utilize arise from the customer's broader ecosystem of journeys being curated by other brands and technologies (most notably, social media). I believe that a customer transformation mindset that adopts a similarly adaptable, holistic view of the customer's interactions within both the marketplace and the larger society and prioritizes consideration of customer needs and goals can break down the existing silos and bring about a more integrated marketplace by bridging the "intervals" or gaps between separate journeys within customers' ecosystems.

In his article, "Why You Need to Map the Customer's Ecosystem," Graham Hill offers an excellent example of the possibilities that arise when we are willing to adopt a more collaborative, inclusive perspective toward the seemingly disparate but actually connected journeys (or what Hill sometimes calls "pathways") in our customers' ecosystems:

> A few years ago, as part of a programme for a Bank client, my team analysed the journey associated with the customer job of 'Finance my Home.' The Bank

was actually interested in the 'Sell a Mortgage' pathway and how it could better sell mortgages. Although they sound similar, they are actually quite different. Much to the Bank's surprise, of the 100 key interactions in the customer's Finance my Home journey, the Bank's Sell a Mortgage pathway only accounted for around 30% of them. The majority of interactions were with a dozen other actors in the customer's ecosystem, including a number whose interests were not aligned with the Bank's, e.g., mortgage brokers who provide independent mortgage advice and comparison platforms who allow customers to easily compare different mortgages.

Of the interactions that were not with the bank, the team identified that an additional 10% easily could have been with the Bank if it looked at the journey through the customer's eyes, rather than at the pathway through its own. It also identified that an additional 30% were with actors that the Bank could potentially have partnered with; that would have been beneficial to the Bank, the partners and to their mutual customers. And finally, that the Bank could have assisted customers with the final 30%, e.g., by providing them with guides, checklists, etc.

The Bank could have increased its engagement with customers from 30% of the interactions during their journey, to 40%, 70%, or even 100%, just by taking a less myopic perspective. Alas, it was only interested in selling mortgages.

With the lessons of this story in mind, Hill advises: "Identify actors that you would like to replace, that might make potential partners for you, or that you can help the customer with. These provide you with new possibilities."

Adopting such a perspective requires the willingness to look beyond how each technology or application you develop for specific customer interactions facilitates only that one step of the journey you directly control and manage for your customers. It requires an understanding that the "customer experience" you're interested in facilitating does not simply start and end with that particular tech or application within your control. And it requires innovative thinking to identify the significant, intersecting moments outside of your brand's own customer journey(s) and to be able to adjust the customer experience to lead to more satisfying outcomes from the customer's point of view. All of this can, admittedly, be challenging. But it is becoming easier with AI's help and increasing partnership openness. More importantly, your customers expect it.

Innovation Collaboration

Ecosystems for business innovation have emerged as a powerful driving force in today's interconnected, technology-driven world. By leveraging cutting-edge technologies such as AI, ML, blockchain, and the IoT, such ecosystems facilitate seamless communication, dynamic collaboration, knowledge sharing, and resource pooling in ways that break down traditional barriers, accelerate innovation, and lead to the creation of novel products and services to meet the ever-evolving needs of customers.

A good example of a relatively mature digital ecosystem is banking, as noted by Kate Brush, a writer for TechTarget. "The ecosystems created through these apps integrate all services and applications into one place, including expense managers, digital wallets, online banking, and digital passbooks." These ecosystems can generate added value to the consumer by incorporating data that go well beyond traditional banking functions. The Danish firm Danske Bank, "created an online system combining customer data with housing-market listings. This provided potential homebuyers with tax, electric, and heating cost estimations; a catalog of Realtors, information, and service providers; and strong, trustworthy financial advice."

The healthcare industry is another sector that profited from early investments in digital ecosystems. "A digital healthcare ecosystem incorporates every touchpoint in a patient's journey, including scheduling appointments, receiving appointment reminders, storing test results, and recording prescriptions." Brush points out that "[m]any healthcare organizations are exploring how to integrate artificial intelligence (AI) and machine learning (ML) into their systems as a way to improve customer experiences and decision-making processes." The digital ecosystem makes this possible by cutting through information silos and "ensuring the correct data is available at the right time, allowing healthcare organizations to take full advantage of the benefits AI and ML offer."

A somewhat more controversial (yet famed) example of an effective digital ecosystem is Apple's, which has been likened to a "walled garden" that can seem off-putting to those on the outside but also praised for the rewards it bestows on those willing to venture in. These include benefits like "the convenience of taking a picture with your iPhone and having it pop up on your Mac moments later or the ability to easily switch your AirPods to the device you're using right now."

APIs and Innovation

A crucial component of modern digital ecosystem

architecture, central to fostering adaptability and innovation, is the Application Programming Interface (API). An API is a set of routines, protocols, and tools that streamline the process of building software applications. By defining how different software elements should interact, APIs offer building blocks that simplify program creation.

Intriguingly, APIs carry a degree of agility that can encapsulate the characteristics of an entire product or service. This means they can be deployed for entirely new applications that the original API developer may not have envisaged. Furthermore, the amalgamation of multiple APIs allows businesses and their collaborators to develop innovative applications, paving the way for unique and creative business models. Consequently, APIs make software development more manageable and open avenues for novel business strategies and new market opportunities across ecosystems.

For instance, an enterprise can establish a digital partner platform via APIs. Consequently, other businesses can avail themselves of these commercial products and services online through the firm's digital assets or third-party partners. The company leverages APIs to reveal its business capabilities to a digital ecosystem, encouraging partners to contribute to the joint creation of financial value.

An example of how APIs can foster innovation is Google Cloud APIs, the "programmatic interfaces to Google Cloud Platform services." According to Google, these APIs allow Google's partners/users to "add the power of everything from computing to networking to storage to machine-learning-based data analysis" to their applications and solve previously intractable problems. The French food industry firm Carrefour as well as the US-based home improvement retailer Home Depot both have touted the impact their use of the Google Cloud Platform has had in accelerating innovation, allowing them, to develop algorithms minimizing the occurrence of "shelfouts" in their stores (i.e., when an item is not visibly available on the shelf for customers to take, even though it may be in stock).

Open Sourcing

Another principle intrinsically linked to digital ecosystems is open sourcing, an innovation methodology specifically tailored to capitalize on the vast potential of collaboration within dynamic platforms. In the contemporary era of globalization and the internet economy, there's an ever-increasing interaction between technology and information. The level of openness and sharing of knowledge, technological advancements, and other scientific and technological innovation results is on the rise. The conventional approach of "closed source innovation" confined within businesses finds adapting to evolving business development and competitive landscapes challenging.

By promoting a transparent and inclusive culture, open-source projects empower individuals and organizations to contribute and build upon the collective intelligence of a global community. This innovation model not only accelerates the development of cutting-edge technologies, it also democratizes access to tools and resources, enabling a diverse array of stakeholders to participate in the process. As a result, open-source innovation has become a driving force behind numerous breakthroughs in fields such as software development, artificial intelligence, and data analytics, fostering a spirit of cooperation and creativity that has the potential to redefine industries and shape the future of technology.

Open sourcing's main concern is the threat it may pose to the profits companies stand to make from their own innovations. However, some have argued that this concern is more than offset by the benefits brought to all parties by accelerated innovation: "Although open-source firms lose the monopoly revenue garnered by closed-source innovation, they gain additional revenue by learning from each other's innovations." To date, open sourcing has been most successful in the software space, although that is beginning to change. Several triumphs have been achieved in data, including the release of open government datasets and initiatives like OpenStreetMap. However, data tied to many commercial machine learning projects often remains a tightly kept secret. Another emerging success story is the open instruction set architecture specification, RISC-V. Unlike previous open hardware projects, RISC-V has employed a unique approach, demonstrating success where earlier projects fell short. While the jury may still be out on the future of open-sourcing, its impact—and the importance of the general trend toward greater transparency and collaboration for driving exponential growth—cannot be denied. As Red Hat researcher Gordon Haff, says, "If you're looking for the next big thing in computing technology, start the search in open-source communities."

Getting Customers Involved

This finally brings us back to the pivotal role that customers themselves may play in a company's commitment to innovation. In the preceding chapter, I stressed the importance of customer-centric innovation in the sense that innovation—just like customer service and a company's business model more generally—must find its North Star in the evolving needs, preferences, and expectations of customers. We will now take this customer focus a step further by considering that innovation can also be customer-*sourced* to a significant degree for companies willing to allow their customers to have a direct role in the innovation process. After all, no one is closer to or feels the impact of customers' needs better than the customers themselves—although they may not always be fully conscious of those needs or understand how best to satisfy them. But with the right nudge and provision of the right tools, customers can often innovate like nobody's business.

Take the case of personalized sneakers, an approach pioneered by Nike in 1999. The company became the trailblazer among athletic shoe brands by offering custom designs online. The [NikeID] service empowered individuals to tailor their shoes to their unique preferences. Customers could select the style, color, material and add text to their shoes through an online application. Other shoe brands have incorporated a customization feature following Nike's pioneering initiative.

Those other brands include big names like Vans, Converse (owned by Nike), and Skors. A 2015 New York Times article on this trend quotes Elizabeth Spaulding, a retail partner at the consulting firm Bain & Company:

> Consumers generally want more control and choice in what they purchase. "There's been a big push toward customization in many industries," Ms. Spaulding said, citing as an example the hugely successful restaurant chain Chipotle, where customers build their own meals. "And when you think about the market for sneakers, there is already a heavy emphasis on choice and individualism."

It's important to recognize, however, that what is involved here is not simply personalization or customization, but *customer* innovation. Indeed, Nike's successor to NikeID, rebranded as Nike By You, refers to it as a "co-creation service," emphasizing the customer's role in the innovation process. Allowing customers to have a hand on the reins of innovation in this way has clearly been one important element in the popularity of Nike's products and has even informed Nike's internal innovation process. As pointed out in this NiceKicks article, "[w]ith NikeID, Nike executives were able to see what 'colorways' ... are popular in the US." From there, they were able to "release specific shoes in specific colorways to areas where they are most popular."

This doesn't mean that every innovation on the part of customers will be a success. One Nike executive acknowledged that in the early days, "some customers designed some really ugly shoes." Nike found, however, that by reducing the options available for customization to a more manageable number, their customers were able to produce more satisfying designs. There is a lesson here: Customers can be highly innovative, but they need user-friendly tools and an appropriate level of guidance to unleash that creativity in ways that will lead to satisfying outcomes for both the company and the customers.

You By Them

I'll close this chapter with a brief question. What if your company had a "[YOUR COMPANY'S NAME] By You" service? As you consider this question, try to look beyond simply offering customization and personalized options to your customers (although that may be a start). Try to discover in which ways you might create a digital ecosystem that would allow your customers to have a direct role in the ongoing innovation of your products and services, whether that be in terms of what is called "exploitative innovation" (i.e., focused on refining and improving existing products and services) or "exploratory innovation" (i.e., the discovery of new opportunities, ideas, and technologies).

And how can you leverage the creative capacity of your customers in ways that don't merely tweak specific channels of customer interaction but take a holistic perspective of the larger "conversation" in which your customers are engaged—a truly "channel-less" perspective? This may involve drawing on the contributions and expertise of numerous external partners for the creation of apps, APIs, and so forth, but don't forget the potential contributions of end users as well. The bottom line: No business is an island, and those businesses capable of tapping into the creative energies of their partners and end users will be better poised to ride the current wave of innovation into the future.

EIGHT

COMMUNITIES AT SCALE

"Alone we can do so little; together we can do so much."

Helen Keller

Gamers, Unite!

In 1997, I found myself caught up in the whirlwind of excitement and innovation that was the early days of online gaming. It was a transformative period, as the gaming landscape had shifted from static content on PCs and consoles to the vast, interconnected realm of the internet. I was fascinated by the seemingly limitless possibilities this new world presented, but I also saw some glaring issues that needed addressing as development teams faced the daunting task of shifting their development, release, and support operations to an online format.

During a typical day of playing my favorite MMORPG (Massive Multiplayer Online Role Playing Game) at the time, I encountered a game bug that wiped out my character's entire inventory. All of my gold, weapons, "loot" that

I had gathered over the last year of playing the game, was gone. I reached out to game support, and the response was short and unhelpful, "Sorry, but there's nothing we can do about it." I was shocked, and outraged. That's when I decided to focus my energy to create a small community centered around one core concept: help make games better.

Initially, our group was a handful of passionate gamers like myself, united by our shared interest in improving the games we loved, spending countless hours debating the finer points of game design, mechanics, and storytelling. As word spread, the community had doubled in size within four months to about 100,000 gamers. Pleased at the growing interest, we considered it a major milestone when, another four months later, we had tripled our numbers. We could have never imagined that at the one-year mark our membership would skyrocket to *one million*.

Our rapid growth and success stemmed from a simple yet powerful idea: *As gamers, we were more than just a number or a revenue source.* We were individuals driven by passion and a desire for change. We sought to create a culture of play that prioritized enjoyment, connection, and growth. Our guiding principles were clear: value our time, invest in quality experiences, and never stop learning, improving, and evolving.

The community continued to thrive and evolve, becoming a beacon for those who believed in its mission. As a community, we were not just gamers but also advocates for change within the industry. Our passion and shared goals allowed us to create a space where players could come together and make their voices heard. And our impact was felt across the industry. Developers and publishers began to take notice of the new ideas and principles we presented. We encouraged these industry leaders to put the player experience first, and we challenged them to excite and amaze us, connect with us, and above all, be honest with us.

I look back with a sense of pride at those wonderful years, when what began as a small, tightly-knit group of gaming enthusiasts became a massive movement that left an indelible mark on the gaming industry. From this experience I discovered *the power of community*, and its lessons apply to any business serious about customer transformation. Simply put, scaling a business by building an invested, engaged community of customers is among the most effective strategies for fostering not only brand loyalty but also market growth and the kind of customer-sourced innovation we discussed in the preceding chapter. When customers are actively engaged and feel connected to your brand, an effectively structured and managed community offers them a platform to nurture relationships and drive the evolution of your products and services—ultimately allowing your business to enhance its bottom line and position in the market.

The Power of Communities

Communities have always been a powerful force in human experience. As Jeffrey Bussgang and Jono Bacon remind us in a 2020 Harvard Business Review article on the competitive advantage of brand communities, "[h]uman beings are fundamentally social animals. Behavioral economics and psychological research have taught us that we fundamentally crave a sense of connectedness, belonging, mission, and meaning . . . Communities deliver these benefits, creating a sense of shared accountability and a set of values while preserving individual autonomy."

In an age of increasingly customized consumer experiences, this need for connectedness is acutely felt by today's customers, who—compared to previous generations—are more empowered and expect to have a say in the development, triumphs, and (potential) demise of the products and services they support. "Product-led growth and bottom-up go-to-market are changing the game, forcing organizations to rethink traditional funnel-focused processes regarding their sales and growth." So says Gareth Wilson, Director of Brand & Content for Orbit. He elaborates:

> Compared to the old world where software tools were imposed upon workers who had little say in the matter, this world is one where purchasing power has become decentralized; where users expect as much from their work products as they do from their personal apps; and where paywalls and 'call us' pricing have been replaced with trials, free tiers, and self-serve plans.

Bussgang and Bacon make a similar observation: "Consumers today . . . expect different relationships with brands. They don't just want a customer support email address and a newsletter; they want deeper interaction with the company and fellow buyers of the product or service."

This trend toward more empowered customers and the concomitant rise of associated communities has been largely organic, attributable to incremental (although often rapid) developments in technology and the general shift toward digitalization. Following the rise of social media in particular, the music industry, for example, quickly gave rise to numerous vibrant fan communities. As María Lucía Villegas of AxiaCore points out, social media platforms "became like an everyday conference for music fans" because social media "allows them to interact with other fans and discuss in real-time their favorite artist." Villegas observes that there are hundreds of digital communities formed by fans "that in revenue translate up to 50% to 80% of the artist's overall revenue," an outcome she notes "may be misinterpreted if we analyze those purchases individually." Villegas points to influencers as another manifestation of the pivotal role of communities in the modern marketplace.

> Influencers know this more than anyone, they need to prioritize their work on building a community, trust, and bond with their followers to stay relevant in the market. Releasing products and profitable content comes second. Most influencers do not become influencers out of nowhere, they can identify what their followers want and provide it to them (adding creativity on how they do that).

Villegas notes that this trend "has evolved into partnerships between brands and creators to collaborate generating content or developing products," an approach she recommends to businesses more broadly. She cautions, however, that such collaboration will only be possible "when brands stop thinking their [social media] followers are their community" (emphasis added). "[A] community means more than the single action of following a brand[;] it implies time and effort put into creating real connections with these humans behind the screens."

Given the current context, Wilson exhorts businesses to have not simply a go-to-market strategy but rather a carefully considered "go-to-community" strategy as well, an emphasis he suggests, "not only helps companies proactively compete in and adapt to this new environment, but also provides the framework and tools to shift from top-down to bottom-up." A community-oriented approach fosters relationships beyond mere transactional exchanges, creating a win-win scenario. More importantly, it elevates the concept of community from being solely company-focused and transactional to a force that influences all facets of the business.

Types of Communities

There are, of course, various types of communities that thrive within the modern marketplace. In virtually all cases, these communities will have a footprint within one or more digital ecosystems, although this does not rule out the possibility of "real-world" elements as well (e.g., special events such as fan conventions held at physical locations). Here is a quick breakdown of the major types of communities, with a few well-known examples:

- Brand Communities: Apple, Google, Starbucks
- Support Communities: Angi, HubSpot, ZenDesk
- Learning Communities: Quora, Reddit, YouTube
- Networking Communities: LinkedIn, Meetup, Indeed
- Social Communities: Facebook, Twitter, Instagram
- Fan Communities: Star Wars, Disney, LA Lakers
- Review Communities: Yelp, TripAdvisor, TrustPilot

The kind of community that your business will want to cultivate depends on who your customers are, the nature of your offerings, and—crucially—your strategic goals, which is to say, the reasons you want to cultivate the community in the first place.

Regarding strategy, it can be tempting to think of community as merely a marketing function (particularly on social media)—the primary marketing goal being to increase brand awareness. However, although there are certainly correlations between community and marketing, the importance of community goes beyond simply boosting brand awareness. Building communities is about developing a deeper two-way conversation (i.e., between your company and those with whom it engages), enhancing the loyalty of all parties to your company's mission, and incorporating community members into the company's operational direction. Because this is a customer transformation book, I'll focus primarily on communities directly involving your customers in the classic sense of the term, but many of the principles covered here can be applied to communities encompassing other external parties (e.g., strategic partnerships) or even internal parties (e.g., your employees).

A Range of Benefits

Let's consider in more detail how your company can benefit from carefully cultivating one or more of the types of communities mentioned above.

Communities foster product improvement and innovation.

In an insightful article on scaling your business through cultivation of an active customer community, Cate Luzio, Founder and CEO of Luminary, suggests that "community is the best market research":

> Your initial customers and early adopters are worth more than any consultant or influencer . . . Listening to your customers and your community will help you with crucial business aspects like product development, pricing strategy, acquisition approach, and more. The feedback is especially valuable in the early days because you're still trying to figure out how everything works.

This is, of course, right in line with the outside-in perspective on the customer relationship that we explored in Chapter 5 and builds on what we discussed in Chapter 7 as well: A key advantage of nurturing a customer community is the opportunity it affords your company to harness your community members' collective wisdom and creativity. Not only will customers help your business innovate by providing valuable feedback on existing offerings, many customers—when they feel their views are valued and when provided an appropriate avenue in which to do so—are eager to share ideas for new products and services. This organic process enables your company to stay relevant, meet customer needs, and develop unique value propositions.

Communities are an efficient way to attract new customers.

As Bussgang and Bacon point out, "Enthusiastic [community] members help acquire new members, resulting in lower customer acquisition costs and a tight viral loop." An active customer community fosters the development of a dynamic ecosystem that expands the reach of your business beyond its initial customer base. The community becomes invaluable for drawing in new customers and building brand awareness through word-of-mouth, referrals, and social sharing. This expansion ultimately results in increased revenue and growth for your business.

Communities generate increased brand loyalty and improved customer retention.

Bussgang and Bacon also note that once you've established an active customer community, "[m]embers are loath to abandon the community, resulting in increased retention and therefore improved lifetime value." Of course, this retentive effect is likely to hold only if the customer community thrives, so your company's focus must not be merely on growing communities numerically but qualitatively as well. (See below for some ideas in this regard.) *Communities improve a business's ability to stay relevant and adapt to changing customer needs.*

Another benefit of maintaining a close, receptive relationship with your customer community is that by doing so your business can be more aware of and thereby be more likely to stay ahead of market trends and shifting customer expectations. This awareness allows you to make proactive adjustments to your products and services to accommodate changing customer preferences, desires, and frustrations, which in turn allows your company to maintain a competitive edge. And more than just the company's products and services, your company's overall mindset and strategic planning is more likely to remain responsive and adaptable if you invest significant resources into fostering a thriving customer community.

Community interaction can enhance and supplement personalized customer service.

In his Forbes article "How to Scale Your Customers and Clients into a Community," Vikram Rajan, a Forbes Coaches Council Member, makes the intriguing observation that active customer communities can help meet the demand for personalized attention that "customers crave" and that has traditionally been delivered by account executives and customer service staff. Thus, "[m]embers support one another, resulting in high gross margins due to a lower cost of service." This mutual support may come in the form of informal technical or product support, tips and suggestions for new or more effective ways to use the brand's products and services, social validation as fellow members of the brand, and so forth. In these ways, community members engage along with the company in the co-creation of service value, yielding higher rates of customer satisfaction at

a potentially lowered cost to your company.
Communities can provide the benefits of networking.
Much has been written about the importance of networking when establishing a successful business or career. Given that a company's community is an extended network, a flourishing community can generate many of the same benefits (both for the members themselves and for the company) traditionally associated with effective networking. Luzio observes, for example, that your community network can have a positive impact on your company's hiring practices:

> When I began hiring, I leaned heavily on my network for the subject-matter expertise needed to identify the right candidates. This network assisted in writing job descriptions, interviewing, and compensation benchmarking for those roles. Without this community, I would have struggled to employ the right candidates and, worse, likely hired incorrectly, which would have cost me both time and money.

Even if your own customer community isn't the sort that can contribute to your hiring process as directly as Luzio's, at the very least a vibrant customer community represents a pool that can be effectively tapped when creatively sourcing job candidates.

Bussgang and Bacon point to the educational firm Codecademy as illustrating the broader networking benefits that can accrue for both community members and the company itself when the community is nurtured and utilized to its fullest:

Since the company was founded nine years ago, more than 50 million people have taken one of its courses. Beyond its rich catalog of interactive educational content, the secret to Codecademy's success has been its ability to link learners who contribute to the catalog and collaborate to improve their skills. Users of Codecademy Pro (the company's paid offering) have access to a Slack group so they can meet, mingle, and share best practices with others and gain access to events with industry professionals and peers. More advanced learners mentor the novices. This rich learning environment generates a network effect in the business model for a company that might not inherently have one.

Scaling a Community

In most cases, vibrant, impassioned communities such as this don't arise independently; they must be cultivated and carefully managed in ways that empower community members and effectively channel their energies. Some of the best tips I've seen for achieving this are presented in a Business Collective article by Tolga Tanriseven, Co-founder and CEO of GirlsAskGuys, a digital community for sharing relationship advice. Here I'll adapt some of Tanriseven's most relevant suggestions and add a few others as well.
Hire the right community managers.

There are certainly technical challenges to address when setting up the best digital ecosystem for your community, but when it comes to actually managing the community, perhaps more important than the technological infrastructure is the presence of community managers who can oversee the human element and be highly responsive to members as people rather than mere numbers. As Tanriseven observes, "[y]ou need people who are passionate, proactive and have people skills, because their role is to manage people rather than content." This isn't to say that you won't supply content—it's important to do so, in fact, so that community members will be exposed to a steady stream of new ideas that spur their thinking and engagement with one another (more on that below). But more important than any content is the empathetic, personalized support provided by community managers. Such support, Tanriseven reminds us, "reinforce[s] the community feeling" and helps prevent members from seeing you as just another "faceless brand entity."

Cultivate a unique community culture.

One of the major benefits of having an enthusiastic customer community is its boost to brand loyalty, and there are few better ways to foster such loyalty than to nurture a community culture that is uniquely yours. In this regard, Tanriseven recommends keeping an eye on the community's language use: "You can foster a unique culture by watching the community language closely and incorporating it into your own communications, as well as by using symbols created organically by the community itself. Reinforce culture by talking to your users the way they talk to each other." A classic example of this is the tightly-knit Harley Davidson community, with its more than 1,400 local chapters. Early on, Harley Davidson wisely embraced the use of "biker language" in many of its promotional materials and within its community. Even the official designation for each chapter, a "H.O.G." (i.e., Harley Owners Group) is based on the slang term "hog," referring to large, heavier Harley-Davidson models. The result of this high level of linguistic awareness is a community with a strong sense of identity and a particularly tenacious brand loyalty.

Encourage content creation that matches your community's interests. In the Forbes article mentioned earlier, Rajan argues that supplying a content-rich platform to your community (with interactive media such as "blogs, podcasts, [and] web/seminars") enables members "to engage with each other and to rave about you to others." Tanriseven expresses a similar sentiment while also emphasizing that the type of content involved must resonate with your members: "Figuring out the types of content that are popular within your community culture can create more engagement among community members." To this end, Tanriseven suggests selecting and showcasing members' own viral content so as "to shape the direction of future content creation." Rajan agrees and recommends encouraging "customers to contribute content (like user-generated content, including use-case photos, case studies and testimonials)."

It's important that your digital ecosystem provide straightforward ways for members to do this. Bussgang and Bacon stress that the process of contributing content should be "simple" and "easily navigable" for members, so that they "can easily create new value for others in the group to consume." This value creation should be "(a) crisply defined, (b) simple and intuitive, and (c) provide almost-immediate gratification."

Maintain open, objective governance of the community.

To maximize the benefits of a customer community, businesses should prioritize open, objective governance and evolution within their core operations. By actively empowering customers to be aware of and even participate in decision-making processes (to at least a limited extent), companies can better ensure that their strategies and offerings align with the needs and values of those they serve.

This inclusive approach fosters trust and loyalty, further strengthening the connection between brand and customers.

Connect your with other communities.

We've already stated that customer communities can bring the benefits of an extended network to both your customers and your company. This also includes building relationships and making connections to other individuals and communities that will potentially benefit your members. Ask yourself: Which social media communities are centered around topics closely related to my products and services? What other individuals, influencers, or brands can bring value to my customer community in ways that complement my brand's value?

Create KPIs to measure and analyze community engagement.

Although I've emphasized the importance of stressing the human element when scaling your community, this doesn't mean you and your community managers should turn a blind eye to the numbers and simply "follow your gut" when making decisions that will impact the community. As with other areas of your business, you'll want to inform your decision-making by identifying your most important KPIs and measuring them consistently. Tanriseven shares that their company, measures

> the number of opinions per user on a daily basis and how long it takes for a question to get a response. These statistics tell us how people are engaging with our site's content and with each other . . . and help us track the overall mood of the community at any given moment (which helps determine content decisions).

To sum up, applying the above strategies when scaling your customer community will ideally lead to synergy between the community and the digital ecosystem in which it is housed. That is, a digital ecosystem that has been carefully designed and managed with a specific community in mind tends to yield an overall more satisfying user experience, fosters creativity and innovation, and facilitates continuous community growth. In turn, the community's engagement and advocacy positively impact the digital ecosystem by driving more traffic, attracting new users, and generating valuable feedback. This self-reinforcing cycle leads to improved digital offerings and a richer, more vibrant community, ultimately resulting in a stronger and more resilient business.

A Challenge Worth Accepting

Growing your business by fostering a thriving customer community is, as Luzio reminds us, a "complex" undertaking that takes a significant amount of time and effort. But, she is convinced the effort is worth it, "For us, listening to customers and sustainably growing our business is paying off in many ways. Grand audiences don't appear overnight, and it is okay for scaling to take a while." By adhering to a community-centric approach, businesses can expand steadily. Numerous successful and highly profitable companies take their time scaling. Still, your company will naturally scale if you prioritize your customer and consistently pay attention to your community's voice.

My own experience leads me to heartily agree. I remember the mantra that guided our groundbreaking gaming community all those decades ago: We are fueled by passion and inspired by change. Building a successful community involves passion, shared goals, and a common perspective. It's about valuing the individuals that make up the community and understanding that their collective strength can drive change and innovation. A thriving community is a testament to this perspective's power and impact on customer experience, loyalty, and transformation.

Case Study: Twitch

The livestreaming platform Twitch is a good case study of a business with community at its core. As chronicled in Stream Scheme, Twitch began in 2007 (at that time Justin. tv) as a livestreaming platform for gamers (and currently remains "the top game-streaming platform in the world"), but it quickly attracted other types of users and over the years has expanded into other areas of livestreaming so that, "[a]s of 2023, users can also stream IRL videos, talk shows, and music festivals." Twitch was purchased by Amazon in 2014 for $970 million and is now a subsidiary of Amazon.

Twitch's rapid growth can be attributed to various factors, but one clear reason is that its user value is generated directly by the community in transparent, accessible ways. Not only can users "learn more about a particular game by watching it played by professional streamers," but "by joining a streamer's online community, they can meet like-minded people who enjoy the same content they do and game with them." Twitch also has "inbuilt socializing services like direct messaging (called whispers) and the ability to chat with others watching the same channel," making it "a great place for gamers to meet and interact with each other."

Even though users have been able to watch most Twitch streams without setting up an account on the platform from the beginning, "many decided to sign up when they saw the benefits of having one." As time passed, the simplicity of initiating a live broadcast across various devices, such as PCs, gaming consoles, and mobiles, attracted an increasing number of Twitch users to start streaming themselves. Coupled with the constant growth of the gaming community, it has propelled Twitch to an imposing stature within the gaming industry.

Twitch has wisely maintained a focus on providing means for community members to generate value for each other, as when "in 2016, Twitch created the "cheering" method, allowing users to purchase Bits . . . and donate them to their favorite streamer." Although the service has its share of casual users, many members of the community "earn a livable wage" from the networking opportunities Twitch offers, making money "in a variety of ways through crowd-sourcing or by working with sponsors. The most common way for broadcasters to earn is through the donations and subscriptions of their followers . . . They can also make money by selling (or trading) their skills to other broadcasters."

Not every business will be able to integrate its offerings so directly into the functioning of its customer community, but Twitch is, an excellent example of what can be achieved when a business prioritizes the needs, expectations, and creativity of its customers and consciously fosters their ability to generate value for each other within a thoughtfully designed and managed community.

Case Study: HPE - A Community Within

The tendency to confine creative efforts to a selected few has ironically led to a talent drain in many companies. Recognizing this issue, certain forward-thinking enterprises are fostering an all-encompassing creativity culture to retain and inspire budding entrepreneurs. A notable example is Hewlett Packard Enterprise (HPE), which has established a groundbreaking program called Idea Matchmaker that drives innovation throughout its organization and builds a thriving internal community of innovators.

Idea Matchmaker, designed by attorney Jeffrey Fougere, provides a platform for HPE employees to share innovative ideas in a globally accessible database. This concept facilitates seamless collaboration and discussion and enables individuals to connect and hone their ideas. As Fougere points out, allowing team members to conceive and fine-tune their ideas collaboratively is empowering.

A distinctive feature of Idea Matchmaker is its automated algorithm, designed to disseminate ideas among employees likely to show interest. Regularly exposing every team member to fresh ideas fosters an environment of perpetual innovation. Creativity authority James Taylor identifies this as a manifestation of the "backstage creativity" trend, where teamwork between colleagues, creative units, and even humans and machines culminates in a more inclusive culture of innovation.

Taylor contests the traditional image of the isolated creative genius, underscoring that creativity is a community effort that thrives on the participation of many. He equates this to a concert where only the performer is visible to the audience while many people work behind the scenes to ensure a successful performance. By harnessing backstage creativity, companies can unlock the latent potential of their entire workforce instead of solely relying on a few industry stars.

The effectiveness of Idea Matchmaker at HPE is due to its unique ability to traverse organizational layers and establish efficient connections between team members. This platform aids in the seamless progression from ideation to testing, approval, and implementation. However, the benefits of this initiative extend beyond mere idea commercialization. HPE gauges the impact of backstage creativity using metrics such as the number of connections forged and ideas viewed. These statistics help cultivate a vibrant company culture, promoting employee engagement and fostering community within the organization.

With rapid technological advancements, creative arenas are becoming increasingly accessible, even to non-engineers. Fougere acknowledges that traditional boundaries are being dismantled by tools like low-code and no-code platforms, which simplify the realization of ideas. HPE intends to leverage large language models like ChatGPT as writing and communication aids to bolster team collaboration, bypass language barriers, and promote the blossoming of ideas.

Businesses can reap significant benefits when they efficiently nurture, execute, and incentivize innovation. Google was one of the first companies to use a peer-to-peer bonus system, allowing colleagues to recognize each other's contributions with small monetary bonuses. This system operates through an internal tool where employees nominate colleagues for bonuses, which managers review and approve, providing insights into team productivity. Google later extended its bonus system to reward external contributors through the Open Source Peer Bonus program. Zappos also has a similar program where employees can reward others with 'Zollars,' an internal currency, for good work, participation, and volunteer work, and there's a $50 co-worker bonus for employees who best represent the company's core values.

In the past, entrepreneurship typically sprang from individuals departing their organizations to pursue more efficient methods or to monetize their creative impulses overlooked at their workplace. But for people who appreciate the collaboration and resources of a larger team, adopting backstage creativity allows them to discover and leverage their inherent talents and skills.

James Taylor advocates for everyone to make the first move toward tapping into their creative potential. By fostering a culture that values backstage creativity, companies can harness the collective strength of their employees, enhance job satisfaction, and unlock new growth and innovation opportunities. In Taylor's words, "Creativity is a team effort. 'Backstage creativity' is about drawing out the best from everyone, not just the industry's superstars." By fostering such an environment, HPE has scaled innovation and built robust internal communities that promote innovation and co-creation, setting an inspiring example for other organizations.

ACTION PLAN

STAGE FOUR: ECOSYSTEM

Introduction
The fourth stage represents a crucial bridge between the customer and the business in the customer transformation process. In this stage, we leverage digital ecosystems to innovate, forge new partnerships, and build relationships that expand into fresh markets. This focus on digital ecosystems and community development lays the groundwork for developing people interfaces through APIs, creating journeys delivered through these interfaces. By opening your products and services to the community, we can dissolve traditional channels, enabling anyone to contribute to your business, driving new ideas, and fostering innovation across various marketplaces, industries, and even amongst competitors.

Day 0: Reflection and Goal
- **Reflection:** Consider how your business currently utilizes digital ecosystems to drive innovation. Reflect on the extent to which you've harnessed APIs to develop people interfaces and how your products and services have been made accessible to

community contributors. Think about the degree to which you operate without traditional channels and the possibilities that can open up when the wider community can contribute to your innovation process.

- **Goal:** The primary aim for this stage is to master the power of digital ecosystems and community engagement as integral components of your customer transformation process. We strive to open your products and services to the broader community, dissolving traditional channels and enabling diverse individuals and entities to contribute to your business. This endeavor will shape your mindset toward seeing the community as a valuable resource for driving innovation and expanding into new markets.

Day 1: Content Walkthrough
We'll explore the power of digital ecosystems and the concept of channel-less innovation. We'll also explore how APIs create journeys, and people interfaces and how a community-centric approach can foster business innovation and expansion.

Day 1: Workshop Questions
1. How can digital ecosystems catalyze innovation within our business?
2. In what ways can APIs enhance customer engagement and transformation?
3. How can opening up our products and services to the community drive innovation?
4. Can we identify potential partnerships that could

expand our business into new markets?
5. What role do communities play in your current go-to-market strategy?

Day 2-7: Homework and Next Steps
Reflect on your current use of digital ecosystems and APIs. Identify potential areas for community engagement and consider how a channel-less approach could lead to broader innovation and market expansion.

Day 8: Touchpoint and Next Steps
1. Share your reflections on your business's current digital ecosystem and API usage.
2. Discuss potential strategies for enhancing community engagement.
3. Identify opportunities for channel-less innovation.
4. Develop an action plan to implement these changes.
5. Schedule the next meeting to review and refine this action plan.

Day 9 and Beyond: Action Plan
30-day plan: Start enhancing your digital ecosystems and API interfaces, focusing on channel-less innovation and community engagement. Identify one service or product to open up to the community to help iterate on.

60-day plan: Implement changes, open your products and services to the community, and measure the impact on innovation and market expansion.

90-day plan: Evaluate the effectiveness of the changes and the impact on customer engagement. Refine your approach based on the results and feedback, and prepare for the next training phase.

STAGE FIVE: CULTURE

Become Inspired by Change

NINE

CULTURES OF PRAISE

"Culture is the key to a successful transformation. Without a culture that encourages change, any improvement initiative will fail."

John Kotter, Harvard Business School

No Time for Introductions

In the previous chapter we emphasized fostering an engaged customer community as a crucial element of our broader emphasis on creating exceptional customer experiences. This customer-centric perspective is already familiar to many organizations, even if they struggle at times to fully execute it. Even fewer of these organizations have dedicated time or emphasis on developing exceptional experiences for their employees even though it is a vital tool for any company serious about customer transformation.

The fact that a motivated, engaged employee community must be actively cultivated often goes unrecognized. In 2019, I was consulting with a large international asset-management company and had arranged to meet in person with members of their team at the company's New York offices to discuss improving their customer-facing projects. While working closely with the team over several weeks to plan an extensive workshop, we identified the key participants, drafted the agenda, and scheduled the three-day workshop involving about 30 team members.

On day one of the workshop, I opened the meeting by introducing myself and my team and outlining the next three days. Since there were more than 30 people in the room, I decided to skip time-consuming introductions and jump right into our first topic.

Barely a minute into my presentation, a gentleman seated at the large conference table raised his hand. He politely requested that we go around the room and do introductions, since he didn't know half the people there. Somewhat taken aback, I asked for clarification: "Aren't you all on the same team, or do we have different teams here?"

Another participant chimed in, "No, we're all on the same team, working on the same project." The original gentleman added, "That's right, but I'm not sure who is who." Trying not to appear too surprised, I agreed to their request, and we proceeded with introductions.

Eventually, one of the executives offered some clarity. "Despite everyone being on the same team, this is the first time we've had everyone in the same room."

I was astounded (and I'm sure by this point it showed on my face). It seemed incredible that a team with a common purpose and goals, responsible for developing customer solutions and curating customer experiences, had never met or even spoken to each other. The incident was a stark reminder that it's easy to become so task-oriented that we forget the human element in the collaborative process and the power there is in bringing people together. This realization helped shape the rest of our workshop, and over the next three days we explored the importance of fostering connections and working together as a team to drive the success of their customer-facing initiatives.

I'd like to do something similar in this chapter and discuss how to foster the kind of employee culture that advances customer transformation. Interestingly, the secret to creating such a culture is to take many of the customer-facing principles we outlined in the preceding chapters and apply them to your employees. For example,

- Customer Empathy = Employee Empathy
- Customer Experiences = Employee Experiences
- Customer Journeys = Employee Journeys

That is, if we leverage the same processes we use to ensure our customers' needs are met, we can build a culture that meets our employees' needs and position that culture to deliver continuous customer transformation. The key principle, put simply, is this: *Treat your employees like your customers.*

In the rest of this chapter, I'll walk you through a methodology I've developed for putting this principle into action: the P.R.A.I.S.E. Method. This methodology outlines the fundamental parallels between the customer and employee experiences that must be managed to effectively support the customer transformation process.

The PRAISE Method

The PRAISE Method is a comprehensive, customer-centric process aimed at helping organizations create a transformative culture by establishing a strong sense of shared Purpose that influences Relationships, Adaptability, Insights, Service, and Empathy. This methodology offers a holistic approach to developing a customer-focused mindset among employees and instilling values that drive positive customer and employee outcomes.

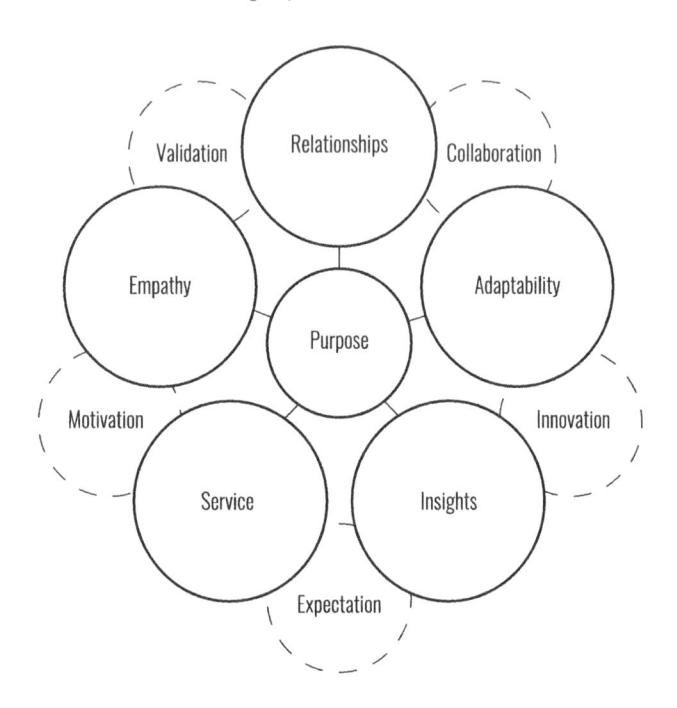

```
                    Relationships
        Validation              Collaboration

     Empathy                          Adaptability
                    Purpose

     Motivation                       Innovation

            Service         Insights

                   Expectation
```

At the heart of the PRAISE Method is *Purpose*, the guiding force behind all organizational actions and decisions. Without a constant, unified sense of a common purpose, your employees will lack clear context with which to fully appreciate and actualize the other elements of the method. The other elements comprise five core pillars (which I'll discuss in greater detail below):
1. **Relationships:** Fostering strong connections with customers and colleagues, enhancing trust, loyalty, and collaboration
2. **Adaptability:** Embracing change and remaining

flexible and open to evolving customer needs and market conditions
3. **Insights:** Leveraging data and customer feedback to make informed decisions and drive continuous improvement
4. **Service:** Providing exceptional customer support and assistance, meeting their needs and expectations
5. **Empathy:** Understanding and sharing the feelings, thoughts, and perspectives of customers and other employees, creating personalized and meaningful experiences

Between these pillars, the PRAISE Method highlights essential integrations:

1. **Collaboration (Relationships + Adaptability):** Encouraging teamwork and cooperation to drive innovation and adapt to changing circumstances
2. **Innovation (Adaptability + Insights):** Continuously seeking ways to improve products, services, and processes to meet customer needs and stay ahead of the competition
3. **Expectation (Insights + Service):** Managing customer expectations by using data-driven insights to deliver tailored experiences and exceed customer desires
4. **Motivation (Service + Empathy):** Inspiring employees to provide outstanding service through genuine care, understanding, and connection
5. **Validation (Empathy + Relationships):** Creating an environment of psychological safety where employees feel heard, valued, and respected

At the Center
Purpose

Purpose drives every action and decision in a customer-centric organization, guiding employees toward a shared vision and common goals. A clear, well-defined purpose inspires employees and resonates with customers, instilling trust and loyalty in both communities. By uniting employees under the umbrella of a shared purpose, organizations answer the ever-present (but often unspoken) employee question, "Why am I here?" A compelling purpose statement ascribes significance and meaning to the many actions and tasks employees tackle. This, in turn, fosters employee commitment and passion, ultimately leading to greater customer satisfaction and the company's long-term success.

To establish a strong sense of purpose, organizations must clearly articulate their mission, values, and objectives, ensuring they are reasonably aligned with the needs and expectations of both their employees and customers. Leaders can then create a more engaged and motivated workforce by consistently communicating the organization's purpose and clearly connecting it to employees' day-to-day tasks.

Incorporating purpose into decision-making and strategy allows businesses to remain focused on the human element, that is, both the company's employees as well as its customers. This is key to developing a powerful, unified employee culture that drives sustainable growth by ensuring exceptional experiences for both the employees themselves as well as the customer community they serve.

The Core Pillars

Relationships – Strong employee relationships are just as vital as strong customer relationships.

Relationships are the foundation of any successful

organization and are the basis for trust, loyalty, and collaboration between employees and customers. Therefore, it is essential to create an environment where employees feel supported and valued, outcomes which improve job satisfaction and performance and translate into better employee experiences across the board. As a survey on Vantage Circle showed, "having a familial relationship with coworkers boosts productivity and feelings of well-being in the workplace."

In a customer-centric organization, these improved employee experiences translate into better customer experiences, and the nurturing of customer relationships should be a top priority. This means understanding customers' needs, preferences, and pain points to provide personalized experiences and solutions that drive customer satisfaction and loyalty. By fostering solid relationships with customers, businesses can establish themselves as trusted partners and create long-lasting connections that drive repeat business and positive word-of-mouth.

There are no easy shortcuts toward building strong relationships: They require a commitment on the part of your organization and its members to teamwork, open communication, active listening, and genuine care and understanding. That said, there are ways your organization can seek to strengthen relationships (both among employees and with customers)—for example, by investing in training programs that develop employees' interpersonal and communication skills and promote a culture of inclusivity and diversity. Team-building activities and events that encourage social interactions (e.g., activities centered around food) are also practical ways to foster relationships that delve below the surface.

By prioritizing relationships, businesses can enhance

their reputation, increase customer and employee retention, and achieve sustainable growth.

Adaptability – A customer-centric organization's employee culture is characterized by a readiness to adapt to rapidly evolving consumer needs and choices.
Adaptability is critical to success in today's rapidly changing business landscape. To maintain a competitive edge, organizations must foster a culture of adaptability that allows them to be agile and flexible, ready to respond to evolving customer needs, market conditions, and technological advancements. (Awareness of the latter is closely associated with what has come to be called "digital acceleration," or the rapid adoption of digital tools as a means to meet continuously evolving customer demands.)

Establishing a culture of competitive adaptability involves regularly reviewing customer feedback, monitoring industry trends, and identifying opportunities to enhance products, services, and processes. However, research has shown that the most challenging aspect of getting an agile transformation right is "the people dimension": achieving employee buy-in. To promote adaptability within an organization, therefore, it is essential to encourage a growth mindset among employees, emphasizing the value of learning from challenges and embracing new ideas. This may involve articulating what has been called the cultural "from-tos," or the transitions needed from old ways of thinking and behaving to new, more adaptable ways. For example, there may be a need in the employee culture to shift from a mindset of "loudest voices winning" to one of "valuing every voice," and from "managing and directing" to "empowering and coaching." In general, providing opportunities for professional development and cross-functional training can help build a more adaptable workforce capable of tackling diverse problems and adapting to new situations. By embracing adaptability in this way, organizations can navigate uncertainty, seize opportunities, and maintain a strong position in the market.

Insights – Data-based insights into the behaviors of both employees and customers are invaluable to informed decision-making in a customer-centric organization. Leveraging customer data has long been a best practice when it comes to identifying trends, uncovering opportunities, and proactively addressing nascent market challenges. The insights thus gathered enable organizations to tailor their products, services, and experiences to meet customers' specific needs, ultimately leading to increased satisfaction and loyalty. In human resources circles, a similar approach to the gathering and analysis of employee data is increasingly common via the growing field of people analytics, the "data-driven and goal-focused method of studying all people processes, functions, challenges, and opportunities at work to elevate these systems and achieve sustainable business success."

While this data-based understanding of employee communities is a welcome development, the insights gained from people analytics must not remain siloed in HR. To foster an employee environment that will support truly sustained customer transformation, it is vital for your business to connect the insights gained into your employees' needs, capabilities, and ways of interacting with each other (and with customers) to the insights you have of your customer community. Sharing these insights across departments fosters collaboration and innovation, as teams can work together to develop new solutions and enhance existing offerings. Doing so also paves the way for a more fully informed practice of data-driven decision making, meaning that decisions can be made "without bias or emotion" and that, as a consequence, "your company's goals and roadmap are based on evidence and patterns you've extracted from it, rather than what you like or dislike." Grounded in a broad understanding of the complementary factors that enrich communities, employees and customers, this objective approach to decision-making offers a more reliable guide to strategic planning and decision-making.

In order to gather meaningful insights into their customers and employees, organizations should implement robust data collection and analysis processes. This includes monitoring interactions, capturing feedback through surveys and reviews, and analyzing transactional and behavioral data. Combining various data sources allows for a comprehensive understanding of both employees and customers, revealing patterns, preferences, and areas for improvement. By investing broadly in analytics tools and data-driven decision-making in this way, organizations can turn insights into actionable strategies that drive growth as well as success. Emphasizing the importance of data-driven insights creates a culture of continuous improvement, ensuring that businesses remain responsive, agile, and customer-focused.

Service – Serve your employees in the same way you serve customers.

Service is a fundamental aspect of a customer-centric organization whose chief aim is to provide exceptional support and assistance to customers, ensuring their needs and expectations are met. A strong commitment to service demonstrates an organization's dedication to creating positive customer experiences and fostering long-term relationships. In the context of leadership, service also implies focus on the well-being and development of employees, promoting a supportive and collaborative work environment. To excel in service to customers, organizations must prioritize their needs, actively listen to feedback, and respond promptly and effectively to inquiries and concerns. This can be achieved by investing in customer service training, implementing efficient systems and processes, and setting clear expectations for employee-customer interaction. By consistently delivering exceptional service, businesses can establish themselves as trusted partners, increasing customer satisfaction and loyalty.

Servant leadership is an essential aspect of creating a service-oriented employee culture. As Dr. Josh Axe states, servant leadership

> is a philosophical concept in which people establish authority not through traditional top-down power structures but by serving with the innate desire to fulfill their team's and community's needs. Those who practice this leadership style multiply leaders by positively influencing those they serve and fostering consistent development in their followers.

Leaders who embrace a service mindset are committed to prioritizing employees' needs by empowering their teams, supporting their growth and development, and leading by example. This approach fosters a sense of unity and shared purpose, ultimately driving the organization to achieve its customer-centric goals.

We'll discuss servant leadership in more detail in Chapter 13. For now, it's worth emphasizing one important effect of servant leadership: By modeling a form of leadership that prizes humility in place of hubris and self-sacrifice instead of selfishness, servant leaders create a safe, positive environment in which employees themselves are empowered to similarly assist and serve both their customers and each other.

Empathy – Drive employee productivity through employee empathy.

Empathy is the ability to understand and share the feelings, thoughts, and perspectives of others, and it plays a crucial role in fostering a customer-centric culture. By cultivating empathy, organizations can create personalized and meaningful experiences for both customers and employees, enhancing satisfaction, trust, and loyalty.

Developing empathy requires active listening, open-mindedness, and the willingness to observe situations from another person's point of view. Empathy helps employees identify customer pain points in a customer-centric organization, tailor their approach to address individual needs, and provide compassionate support. Similarly, employees who exercise empathy toward colleagues tend to be more mutually supportive both in facing challenges and celebrating success in the work environment. Moreover, research has shown that "empathy in the workplace is positively related to job performance." (Conversely, a lack of empathy can have just the opposite effect.) Generally speaking, by demonstrating empathy, businesses can better connect with their employees and customers, fostering deeper relationships and driving long-term success.

In order to promote empathy within an organization, it is essential to create an inclusive and diverse environment where individuals feel heard and valued. Encouraging open communication and genuine perspective-taking, cultivating compassion, providing training in empathy and emotional intelligence, and recognizing empathetic behaviors can all help foster an empathetic culture.

The Integrations

Collaboration (Relationships + Adaptability) – Harness the power of collaboration to generate positive employee and customer experiences.

A collaborative work environment is essential in a customer-centric organization, as it fosters open communication, knowledge sharing, and collective problem-solving. Collaboration requires employees to work together towards common goals, leveraging their unique skills, perspectives, and experiences in order to achieve better outcomes. By promoting a collaborative culture, organizations can create more effective solutions, adapt more quickly to changing market conditions, and enhance the overall customer experience.

When seeking to foster a collaborative environment, businesses should prioritize open communication, inclusivity, and teamwork. Encouraging regular feedback and idea sharing will help create an atmosphere where employees are engaged and feel their ideas are valued. Promoting ownership of ideas and rewarding evidence-based risk-taking will provide increased motivation for working together to reach solutions. Providing cross-functional collaboration opportunities, such as team-building exercises, workshops, and group projects, will help break down silos and create a more cohesive organization. Implementing technology that supports collaboration, such as project management tools, communication platforms, and shared workspaces, will further enhance the collaborative process. By promoting a culture of collaboration, organizations can harness the collective intelligence of their workforce, driving continuous improvement in the process of customer transformation.

Innovation (Adaptability + Insights) – Offer every employee the motivation and resources to innovate.

In a customer-centric organization, establishing a culture of innovation is critical for staying ahead of customer needs, adapting to market changes, ensuring customer satisfaction, and maintaining a competitive advantage. To promote innovation, it is essential to create an environment within an organization that encourages creativity, experimentation, and learning from failure. This involves providing employees with the resources, time, and autonomy to explore new ideas and pursue novel solutions. Recognizing and rewarding innovative thinking will help reinforce the importance of innovation and inspire employees to think outside the box.

It's also important to remember that innovation is not achieved by a few creatives working in isolation. As a Washington Post article observes, the solution to the challenge of "innovating at speed and scale" is collaboration: "Collaboration allows organizations to bring their best thinking to bear on a problem, and it's the wellspring of invention." Dominic Price, a Resident Work Futurist for Atlassian, adds that "[t]he defining feature of an innovation culture is the belief that innovation is every employee's job, not just the domain of a few." Collaboration allows employees to share ideas, knowledge, and insights, leading to more effective problem-solving and creative solutions. By embracing organization-wide innovation in this way, customer-centric organizations can stay agile and responsive, continually evolving to meet the changing needs of their customers.

Expectation (Insights + Service) – Maintain expectations that bring out the best in (and for) your employees.

Expectations are crucial in customer-centric organizations: Customer and employee experiences alike tend to be impacted positively when expectations are met and negatively when they are not. For this reason, it is imperative that stakeholders' expectations are aligned with what the organization and its employees can provide. Setting and managing expectations involves communicating goals, standards, and desired outcomes to ensure such alignment.

In the context of customer experiences, understanding and meeting customer expectations is key to delivering value and creating positive interactions. This requires organizations to actively listen to customer feedback, monitor industry trends, and develop a deep understanding of their customers' needs and preferences. By delivering on promises and consistently meeting or exceeding expectations, businesses will foster trust and build a reputation for reliability and excellence.

The setting of clear employee expectations is also essential and tied to providing direction, motivation, and a framework for performance evaluation. However, it is crucial that employee goals and expectations align with the fundamental principle stated earlier: in a customer-centric organization employees should, in a real sense, be valued in ways similar to the ways businesses have traditionally valued customers. This means that the ultimate aim of setting expectations and measuring performance isn't just about profits, but ensuring a good employee experience. This in turn directly affects the business's relationship with its customers. Indeed.com reminds us, "Happy employees are more likely to create a positive experience for your customers or clients. You can tell when an employee likes their job and is a fan of their company. Customers pick up on that." With this in mind, businesses will benefit from the creation of a shared sense of positive purpose, encouraging employees to strive for excellence in their work while simultaneously demonstrating to employees that they are appreciated and valued.

Motivation (Service + Empathy) – Prioritize employee motivation to create a passionate, driven workforce.

Employees remain engaged, committed, and inspired to deliver exceptional customer experiences when they are sufficiently motivated to do so. Organizations can, therefore, increase productivity, satisfaction, and overall performance by fostering a culture that supports and rewards motivation.

To achieve this, employees need a clear sense of purpose and the ability to see how their work connects to the organization's mission and customer-centric values. It is also essential to provide personal and professional growth opportunities, such as training, mentorship, and career development, which will help employees feel more capable and invested in their roles. Recognizing and rewarding employees for their achievements and contributions to customer success can further enhance motivation and reinforce the importance of customer-centricity while driving continuous improvement. Lastly, establishing a supportive work environment that encourages open communication, collaboration, and the uninhibited generation of ideas is a prerequisite for ensuring a fully engaged workforce.

Validation (Empathy + Relationships) – Validate your employees to bring out the best in them.

Individuals feel validated when their feelings, thoughts, and experiences are acknowledged and affirmed. Valuing the perspectives of customers and employees in this way creates an environment of trust, respect, and understanding, ultimately leading to stronger relationships and better customer and employee experiences.

During customer interactions, validation involves actively listening to customer feedback, empathizing with their concerns, and demonstrating a genuine commitment to addressing their needs. This approach helps build trust and rapport and generates valuable insights for continuous improvement of the customer experience.

Validating employee experiences is equally important, as it creates a safe, supportive work environment where individuals feel heard, valued, respected, and free to challenge the ideas of others. When employees feel "psychologically safe," they are free to communicate more openly and "share their ideas and concerns without being mocked or humiliated," ultimately leading to greater productivity and gains for the organization. Indeed, research has shown that employees who feel psychologically safe "bring in more revenue, and they're rated as effective twice as often by executives." Validating your employees can begin with something as simple as a "thank you" for a job well done, but also something more tangible such as flexibility in scheduling, appropriate compensation, and continuing professional development.

The Bottom Line

When we show our employees the same level of respect and appreciation that we extend to our customers, the impact on the company's culture, performance, and overall success can be dramatic. The PRAISE Method serves as a concrete framework for addressing this often overlooked element in our strategizing for customer transformation.

TEN

DELIVERING INSPIRATION

"We keep moving forward, opening new doors, and doing new things, because we're curious, and curiosity keeps leading us down new paths."

Walt Disney

A Magical Kingdom

Disney has always been a staple in my life. One of my earliest memories is my family's pilgrimage to Walt Disney World in Florida to celebrate my seventh birthday. I remember the excitement of piling into our old green station wagon for the long trip and the thrill of seeing the magnificent park gates during our final approach. It all felt magical.

And then, just as we pulled into the parking lot, the sky seemed to fall, and within a matter of seconds, the only thing we could see through the car windows was a sheet of relentless rain—a typical Florida thunderstorm.

I remember my parents exchanging glances and saying, "There's no way we're spending money to go in when it's pouring like this."

I was heartbroken. As tears began to well up in my eyes, I pleaded, "Please, let's just go in!"

And at that moment (just like a scene from a Disney animated movie), the clouds parted, the sun began to shine through, and the downpour became a trickle until it ceased entirely. As I looked out with astonishment, a wave of joy came over me, as if some divine intervention had cleared the weather to allow me to enjoy my special day.

Over the years, my respect for Walt Disney and his vision has deepened. My passion for Disney as a brand has been challenging at times, but the theme parks have always been a case study of inspiration for me. The thrilling adventures, delectable treats, whimsical characters, and live shows all symbolize an ever-evolving canvas that, in its unique way, pays homage to Walt's original vision, "Disneyland will never be completed. It will continue to grow as long as there is imagination left in the world."

My family, friends, and I have made countless trips to Disneyland during the past 30 years. I've witnessed a ton of transformation during those times. Of course, there have also been some changes that didn't exactly ring right. Sadly, some alterations to the park have felt entirely profit or political-driven, devoid of the magic that makes Disneyland unique among other theme parks. However, underneath the business of a theme park is still Walt's ideals of imagination.

During one visit a few years ago, I was chatting about the park's many transformations with a friend who voiced their concerns. While I largely agreed with their view, I looked for clarity with, "Don't you think the changes are inspirational?"

I've often wandered through the park, daydreaming and soaking in the sights with a sense of awe. Later that day, reflecting on my earlier question, an idea began to take shape. As I walked around the new construction, renovations, and crowd of happy faces, I felt that familiar tingle of inspiration wash over me. I saw a child delighting in an ice cream cone, a girl grinning at her bubble blower, another brimming with anticipation for a princess autograph, and a couple capturing their happiness with a selfie. It then dawned on me that just as it was good to anticipate with wonder the many changes that would inevitably come to Disneyland—even knowing that some of the changes might seem less magical than others—in the same way, I didn't need to fear the future of my life and of my business with the inevitable changes those journeys entail. Instead, I needed to embrace the changes and draw inspiration from them. From this realization, a motto formed: "Fueled by Passion. Inspired by Change."

When I established my digital agency, this motto became our guiding principle, and it drove us toward innovation and service delivery that consistently inspire our clients. The key, I realized, is to accept that change is inevitable. Technology will continue to evolve, business cultures will transform, and customer expectations and needs will shift—sometimes unexpectedly. But by recognizing the constancy of change, we become free to adapt to it—even be inspired by it—rather than shy away from it.

The preceding chapter emphasized cultivating great employee community experiences in order to feel valued and empowered toward effective customer interaction . In this chapter, I'll revisit and dive deeper into one of the "core pillars" of a thriving employee community presented in the preceding chapter: *adaptability*. Specifically, I propose that fostering an adaptable employee culture involves helping employees be *inspired by change* so that they learn to see change as an opportunity rather than a threat.

Of course, this perspective rubs against human nature: We tend to resist change because it is inconvenient and threatening. Sometimes we even downplay the extent of the change or disregard it rather than facing its challenges head-on (including some business executives who should know better). But this head-in-the-sand approach is unrealistic, especially in today's world. In chapter 1, we emphasized the rapidity with which technology, the marketplace, and customers are changing, and stressed that any successful customer transformation strategy must adapt to this reality. The best way to achieve this is by being motivated rather than threatened by change, to use change as a springboard for delivering inspiring services to your customers.

Case Studies in Inspired Agility

Let's consider a few classic examples of companies that have done just that: They deliberately used inevitable change as a catalyst for inspiration.

Spotify

As the world's leading music streaming service with more than 400 million subscribers, Spotify has become a famous case study for its agile transformation. At a time when music piracy appeared to have struck a death knell in the music industry, Spotify entered the arena and "made legal streaming so easy that, combined with the music industry's (at times heavy-handed) consumer awareness campaigns, new generations were taught to pay for music again."

Although Spotify's achievements are undoubtedly (to some extent) attributable to its innovative offerings and savvy marketing moves, the company's sustainable business model has also been a major factor in its success. Central to Spotify's approach to change has been its adoption of an agile model fueled by an organization-wide culture of agility and customer centricity that features a decentralized decision-making process: "The Spotify model champions team autonomy, so that each team (or squad) selects their framework (e.g. scrum, kanban, scrumban, etc.) Squads are organized into tribes and guilds to help keep people aligned and cross-pollinate knowledge." This model has enabled teams to rapidly iterate and respond to customer needs, improving the overall user experience.

Of course, the question of decentralization versus centralization of decision-making has been long debated, and there are pros and cons to leveraging team-level autonomy as extensively as Spotify has over the last decade. However, Spotify's openness to change (including ongoing business model changes) continues to this day, and research has demonstrated the overall effectiveness of Spotify's approach. The research reveals that Spotify's practice of scaled autonomy does not lead to disorder or unrestricted leniency. Instead, it places the onus on the squads to be responsible for their work, collaborate, communicate, and synchronize their actions with others, all within a framework of certain enabling restrictions. Furthermore, the study found that squads often make numerous decisions independently, devoid of managerial control, primarily due to joint efforts that surpass conventional organizational boundaries. The researchers identify many "mechanisms and strategies that enable self-organization" of this scale at Spotify, including "effective sharing of the codebase, achieving alignment, networking and knowledge sharing."

Lessons from Spotify's efforts to build a customer-first culture include:

- significant impact of culture on employee satisfaction and customer focus
- effectiveness of Spotify's decentralized approach (i.e., autonomous squads, tribes, and guilds) for stimulating inspiration and customer-centered delivery

ING Bank
Inspired by Spotify and other digital leaders, ING, a Dutch multinational banking and financial services corporation, began a massive agile transformation in 2015 to better align employee efforts with customer needs. To achieve this, (similar to Spotify) the bank decentralized decision-making by adhering to the "end-to-end principle" and working in multidisciplinary teams (or squads) that comprise a mix of marketing specialists, product and commercial specialists, user-experience designers, data analysts, and IT engineers—all focused on solving the client's needs and united by a common definition of success.

These squads are grouped into "tribes" that focus on specific customer journeys and are empowered to make decisions and rapidly innovate. The result? ING has improved time to market, boosted employee engagement, and increased productivity.

According to Global SME/MC Digital Platform Centre Director Adam Walendziewski at ING Bank Slaski in Warsaw, Poland, the most challenging aspect of ING's agile transformation was building an agile mindset within the bank's employee community:

> We dedicated much time to ensuring everyone understood why the transformation occurred. People would only commit to working differently if they could see

why a change was needed and if they could visualize for themselves a future where the delivery would be smoother and more beneficial for all concerned. We knew that without a shared mindset, everything would collapse.

Bart Schlatmann, former chief operating officer of ING Netherlands, similarly assessed culture as "perhaps the most important element of this sort of change effort." He summarizes ING's strategy: "We have spent an enormous amount of energy and leadership time trying to role model the behavior—ownership, empowerment, customer centricity—that is appropriate in an agile [organization]. Culture must be reflected and rooted in everything we undertake as an organization and individuals."

Takeaways from ING Bank's experience include the following:

- effectiveness of ING Bank's decentralized tribes in fostering inspiration and collaboration
- importance of employees having a clear understanding of the company's vision for change

LEGO

In 2015, Danish toy company LEGO was named by Brand Finance as "the world's most powerful brand." It is currently the world's largest toy company, with a 2022 market value of more than US$9 billion. LEGO's current success is all the more stunning because it follows the company's near demise in the early 2000s when costs far outstripped revenue due to a series of misguided design decisions. Spooked by the rising popularity of computer games in the 1990s, LEGO had "let its design team run wild" in an effort to make flashier LEGO sets. As a result, "the number of uniquely-designed bricks went from 7,000 to 12,400, causing production costs to increase." Demand, however, "didn't follow suit," and by 2003, the company hit its low point, "mired in debt to the tune of US$800 million and facing bankruptcy." Fortunately for LEGO, the crisis motivated company executives to implement a more collaborative decision-making process that eventually pared down the number of bricks back to 7,000 and spurred the company's agile journey. Over time, LEGO adopted a Scrum framework to organize its product development and software engineering teams. This shift led to an increased focus on collaboration, continuous improvement, and rapid delivery of value to customers.

One feature of LEGO's remarkable recovery was a conscious decision—pushed by new CEO Vig Knudstorp—to begin listening less to "experts" and more to its core audience: children. Enlisting "actual kids to help invent new products," LEGO became more customer-centric and even opened up a previously unexplored but highly motivated market in the form of AFOLs or Adult Fans of LEGOs. "AFOLs were given their platform to submit ideas and vote on new products, many of which became very successful. Nostalgia among parents led to sets themed around Home Alone, Ghostbusters, and even "The Golden Girls."

LEGO's recovery and rise as a highly successful, agile corporation present several crucial reminders, including:

- importance of customer-driven innovation in digitally-driven world
- power of adaptation in darkest times
- ever-present potential for finding new ways to in-

spire and expand one's customer base

Amazon

Amazon's culture has long been famous for its customer obsession and agility. One way Amazon addressed this problem was to restructure the way teams were organized so as to "maximize their ability to stay close to customers and their needs" and "rapidly launch innovative products and services on their behalf." The most colorful expression of this reorganization was the "two-pizza rule," which states that each team should be small enough to be fed by just two pizzas: "Ideally, this is a team of fewer than 10 people . . . smaller teams minimize lines of communication and decrease the overhead of bureaucracy and decision-making. This allows two-pizza teams to focus more on their customers and constantly experiment and innovate on their behalf."

As Daniel Slater, Worldwide Head of Amazon Web Services' Culture of Innovation, explains, the initial success of the two-pizza-team concept was not as much about numbers per se as it was about "fostering and pushing ownership and independence down to the team level—from ideation to execution, from ongoing operational improvement to constant product iteration and innovation." The associated benefits were many: The smaller team size allowed teams "to run fast, experiment early and frequently, and apply learnings rapidly to drive value to their customers constantly."

Ultimately, however, the two-pizza rule proved insufficient. In time, Amazon discovered that "the biggest predictor of a team's success wasn't whether it was small but whether it had a leader with 'the appropriate skills, authority, and experience to staff and manage a team whose *sole* focus was to get the job done.'" Small teams were most successful when "one highly skilled person was put in charge—and not only had the authority to see the project through . . . but was allowed to focus solely on seeing the project through." Teams with such leaders came to be known as "single-threaded leader (STL) teams, a term borrowed from computer science that means to only work on one thing at a time."

From Amazon's efforts to remain agile in the wake of massive growth, we can glean the following:

- the role of team size in promoting agility, where smaller, dedicated teams tend to be better able to stay focused on customer needs and more rapidly respond to them
- the importance of task focus, both for the team itself and for its leader, who must be capable and empowered to see the project through to completion relentlessly

The Challenge Is Worth It

These case studies illustrate that transformation to a more agile, adaptable business model can certainly be challenging, but it is also within reach. On this note, it's interesting that of the six main challenges to agile transformation listed in a Clearbridge Mobile article, the first three all have to do with mindset:
1. "Failing to communicate vision and strategy for an agile transformation."
2. "Failing to get buy-in."
3. "Failing to adopt a new organizational culture."

So much rides on your organization's attitude toward change, depending on whether you view change as a threat or an opportunity to inspire.

I am reminded again of how the example of Disney spurred my appreciation of an agile mindset. This corporation has historically placed imagination and inspiration at the very heart of its brand. Of its "Imagineers" (i.e., the creatives at the forefront of Disney's design innovations), Disney proclaims, "one of our greatest rewards is knowing that the experiences and environments we create inspire others." Brand marketing aside, this sentiment is actually quite powerful, and it's a worthy goal for virtually any type of business. Inspire your employees to view change—in technology, the marketplace, and customers—as a catalyst for developing more satisfying services and products. When you foster such an employee culture, you'll find you have opened a pathway to inspiring your customers and the broader society.

ACTION PLAN

STAGE FIVE: CULTURE

Introduction

The fifth stage involves transforming your company culture to better align with your customers. The premise is straightforward: shouldn't you do the same for your employees if you're committed to understanding and delivering exceptional customer service? By mirroring your relationship with your customers to that with your employees, your business will become stronger and more profitable. This session features the "PRAISE" method (Purpose, Relationships, Adaptability, Insights, Service, Empathy) to foster cultural transformation that reflects onto your customers.

Day 0: Reflection and Goal
- **Reflection:** Examine your current company culture and evaluate how closely it aligns with the needs and values of your customers. Consider the current state of your relationship with employees and how it mirrors your relationship with customers. Reflect on the degree to which your organization embodies the principles of the PRAISE method.

- **Goal:** The primary objective of this stage is to initiate a transformative shift in your company culture that echoes onto your customers. We aim to adopt a culture that values the same principles you want to extend to your customers - principles embodied in the PRAISE method. The aspiration is to create a work environment where the same level of care, understanding, and dedication is given to customers is also extended to employees. The mindset we want to foster is one of mirroring - by aligning the treatment of your employees to that of your customers, we aim to build a more robust, customer-aligned, and profitable business.

Day 1: Content Walkthrough
We'll explore the concept of cultural transformation and how it links to customer transformation. We'll also delve into the PRAISE method and how implementing it in your company's culture can improve customer service.

Day 1: Workshop Questions
1. How can the PRAISE method be implemented in our company's culture?
2. How does a positive relationship with employees reflect on customer service?
3. Can we identify areas where our company culture needs to be more adaptable?
4. What insights can we gain from our current company culture that will improve customer service?
5. How does empathy play a role in transforming our company's culture?

Day 2-7: Homework and Next Steps

Reflect on your company's culture and how it reflects on your relationship with customers. Identify areas where the PRAISE method could be implemented and start thinking about how to make these changes.

Day 8: Touchpoint and Next Steps

1. Share your reflections on your company's culture and its relationship with customers.
2. Discuss potential strategies for implementing the PRAISE method.
3. Identify opportunities to improve adaptability within your company culture.
4. Develop an action plan to transform your company's culture.
5. Schedule the next meeting to review and refine this action plan.

Day 9 and Beyond: Action Plan

30-day plan: Implement the PRAISE method in your company's culture and transform how it interacts with its employees and customers.

60-day plan: Continue to roll out changes and measure the impact on your employees' engagement and customer satisfaction. Measure the level of adaptability demonstrated by employees and how your organization recognizes talent in the same way you notice customers.

90-day plan: Evaluate the effectiveness of the transformation and the impact on your business. Refine your approach based on the results and feedback, and prepare for the next training phase.

STAGE SIX: TECHNOLOGY

Accelerate Data-Driven Purpose

ELEVEN

PURPOSE-DRIVEN TECHNOLOGY

"The technology you use impresses no one. The experience you create with it is everything."

Sean Gerety, The Home Depot

A Failed Uprising

On Shark Tank's first episode of its thirteenth season, entrepreneurs Kristen and William Schumacher made the case for the Sharks to invest $500,000 (a 3% stake) in their alternative bread company, Uprising Food. By most accounts, Uprising's low-carb, fiber-rich, gluten-free, and dairy-free bread was surprisingly delicious, and the pair came with impressive numbers to back up their request: "During Uprising Food's first year of business in 2019, it earned $85,000. The company's sales totaled about $1 million in its second year. Six months before appearing on "Shark Tank," Uprising Food made $2.1 million, double the prior year's revenue," (Tran, 2023).

And yet, the Sharks didn't bite. One by one, every Shark dropped out. Barbara Corcoran claimed she would lose sleep by investing in Uprising Food, deeming the entrepreneurs financially untrustworthy. Lori Greiner, Kevin O'Leary, and Guest Shark Emma Grede quickly followed. Mark Cuban, the last to exit negotiations, said he didn't understand how the company wasn't more profitable and seemed particularly perturbed when William Schumacher favored marketing jargon over a direct answer.

Nevertheless, following their appearance on the show, it seemed the Schumachers might end up having the last laugh. Shortly after the episode's October 2021 air date, the couple boasted that their company had received "an overwhelmingly positive response" from the public after their Shark Tank appearance. In a December 2021 interview, the husband-wife team attributed their continuing success to such factors as maintaining "uncompromising quality in all things," creating a "family culture" with their customers, and having a mission "bigger" than themselves, namely, to "create a national movement . . . to bring the healthiest staple products to all people and create a mission that can forever live on."

This admittedly noble higher cause notwithstanding, the company permanently went out of business on March 23, 2023. Although no public explanation was offered, it seems that Uprising's demise boiled down to two things: First, the owners did not understand the market or their customer base as well as they thought. Consider, for example, that from the start, they were charging $12 for a loaf of bread, which had risen to $15 by February 2023. With such prices, any company would be hard-pressed to spur and sustain a "national movement." Second, they appear to have invested more in the technology to produce and advertise their product than they did to connect their purpose to their customers. Few people will join a self-proclaimed movement if they cannot see a clear, genuinely achievable outcome.

Recall in the first chapter of this book, I proposed the following principle:

You must be able to connect every technology decision you make, back to one or more specific, documentable customer value propositions. Always.

I emphasized this, in part, as an antidote against the all-too-common tendency to pursue technological and digital advancement for its own sake, as if more and better tech will assure business success. Instead, your tech advancements should be purpose-driven in ways that generate genuine value. However, I want to caution that when we speak of a "purpose-driven brand" or "purpose-driven technology," it's not enough to have in view even the noblest "higher" purpose or cause, one that—if achieved—would genuinely make the world a better place (e.g., everyone having access to nourishing, healthy bread), but not if the attainment of that purpose or cause isn't grounded in the valid needs and wants of customers. The closure of Uprising Food, unfortunately, demonstrates this.

What Sorts of Purpose?

To be clear, I'm a big supporter of businesses seeking to offer high-quality, sustainable services and products to improve the human condition and promote a more healthy, happy, and equitable society. There are many significant causes worth supporting, whether it's doing more to safeguard the environment, promote more inclusive social networks, or foster improved physical and mental health among the masses. And there are clear business benefits to being purpose-driven in this sense: As Lember

Gordon, Head of Marketing at American Golf, Europe's largest golf retailer, says, "Consumers expect companies to give back as standard . . . Doing what it says on the tin is no longer enough. Shoppers want to support brands with a conscience because it makes them feel better."

Practically speaking, however, not every business is the sort whose primary offerings can directly contribute to pursuing such goals. Yet, some observers seem to assume or imply that these are the only "purposes" that should be referenced by the phrase "purpose-driven."

Let's consider a broader, more realistic approach. If your business can find ways to promote weighty societal or even globally-relevant causes directly, do so. Your employees and customers will likely thank you; the empathy, care, and "heart" you show will enrich your brand; and you'll promote positive change in the world. Without diminishing the value of meaningful impact on a societal/global level, there seem to be at least two other senses of "purpose" that are legitimate ends for a business to pursue.

One is the ethical pursuit of *profit*. Although this may seem to lie at the other end of the spectrum from the lofty goals just described, the pursuit of profit can be as legitimate and, indeed, as necessary as the pursuit of a more equitable, humane, wholesome world. The use of the word "ethical" is quite intentional here, as we're all familiar with instances in which the raw, unchecked pursuit of profit has led to various degrees of adverse outcomes ranging from the mildly disturbing to the outright horrific. What, then, is it that can make the pursuit of profit ethical and even wholesome in its own right?

To answer that question, let's consider a third type of purpose that is in some ways more fundamental than either of the above and can offer a healthy check on and supply meaning to the pursuit of profit. It can also ground the pursuit of a better world in attainable objectives that address the immediate needs of relevant stakeholders. Simply put, this third purpose is people—the individual customers, employees, and partners with which your business interacts and for whom it operates.

Consider: Even when there is a higher cause to support society and humanity, you still need people to execute it and customers to buy into it. No matter how strongly you believe in a cause, you can't successfully support it if you can't convince your customers it's where they should be. Similarly, no matter how much profit you may make, it rings hollow if that profit and the course you followed to obtain it haven't enriched the lives of the people you care about and the many customers and stakeholders who helped you get to that point.

Whether you're thinking of the traditional "business" purpose of turning a profit, growing your business's market share and expanding your business, or the more recent ethical emphasis on making a positive societal impact (in the commonly used sense of being "purpose-driven"), the common denominator is *people*—all those who interact with, partner with, work for, and (crucially) buy from and hopefully promote your company to others. As a RevBoss article says, "Being purpose-driven comes down to companies finding the 'why' behind their existence and emphasizing how their employees, customers, and the world benefit from it." By placing people—and in particular our customers—front and center in this way, we safeguard ourselves against "using" our customers and other stakeholders merely as means to our ends, whether those ends are positively changing society or improving our own as well as our shareholders' bottom lines. Keep in mind that both outcomes are acceptable. *Of course* we should want to help build a better world, and *of course* we would want to make a profit while doing so. But as the famed philosopher Immanuel Kant pointed out centuries ago, people are not merely means to our ends; they are *ends in themselves*, and we cannot ethically and humanely relate to them—in business or otherwise—if we don't recognize and respect them as such.

Good Experiences vs. Ethical Experiences

While we're considering a business's ethical obligations to its customers, there's an important distinction to be made between wanting to craft "good" experiences for our customers versus fostering "ethical" experiences for them. This reminds me of something TikTok CEO Shou Zi Chew said before the House Energy and Commerce Committee in March of 2023 to address concerns about the popular social media app's data collection practices. During testimony, Chew made the following comment: "In principle, we want to provide a good experience for our users in general." On the surface, this statement may seem unobjectionable. When we consider, however, that businesses often consider a "good" customer experience one that increases sales and revenue, pursuing such experiences may reflect a prioritization of the business's interests over ethical consideration of the customer.

A typical illustration of the distinction between good and ethical experiences is the use of persuasive marketing and advertising techniques. Businesses frequently employ emotional appeals, scarcity, and social proof to entice customers to purchase products or services. Although these tactics can effectively increase sales, they can sometimes be manipulative and unethical, especially when exploiting customers' vulnerabilities or compromising their autonomy (a topic we explored at some length in Chapter 4). For instance, a business may use emotional appeals to persuade customers to buy a product they do not need or would otherwise desire. While this may, in some sense, create a positive experience for the customer (the thrill of making an impulse purchase, etc.), it is ethically questionable as it capitalizes on their emotional state, hindering their ability to make informed, financially responsible decisions. Another factor to consider is the company's treatment of employees. Some businesses offer low prices or quick service but at the expense of fair wages or safe working conditions—an unethical tradeoff. The use of technology introduces a wide range of ethical considerations (also touched on in Chapter 4), including the potential for misuse of personal information and the risks associated with misinformation, deep fakes, unregulated AI, and autonomous technology, to name a few.

The key to ensuring ethical customer experiences is genuinely empathizing with customers by understanding and prioritizing their needs and interests. This requires active customer engagement, soliciting feedback, and gaining insights into their preferences, concerns, and challenges. This valuable information can then inform the design and development of products and services that meet customers' needs and demonstrate genuine care and understanding of their circumstances. Honest and transparent communication is also crucial, including being upfront about potential product and service limitations, risks, or adverse outcomes.

Sometimes prioritizing ethical over merely "good" customer experiences requires making difficult decisions that take a long-term view of the business's interests, with an emphasis on building lasting, productive customer relationships. Today's customers can quickly identify companies that do not prioritize the customers' best interests, and companies that empathize can differentiate themselves from competitors and create a unique value proposition that appeals to their target audience more ethically.

Purpose Accelerates Your Mission

With this ethically-mindful, people-focused, customer-centric approach to "purpose-driven" businesses and technology in view, we have the context in place to consider one of the primary business benefits that come from prioritizing the customer as the guiding principle for all technological decisions and advancements: doing so can dramatically accelerate your business mission. Purpose-driven technology is closely associated with *digital acceleration* in this regard, and the relationship between the two is profound and intertwined. Let me explain.

Digital acceleration refers to the rate at which digital technologies evolve and are adopted. It has increased significantly in recent years due to factors such as the COVID-19 pandemic, which forced businesses and individuals to rely more heavily on digital solutions. Purpose-driven companies often leverage this digital acceleration to drive their mission and create positive impacts. Notably, the faster businesses can adapt and leverage technologies that fulfill a clear, concrete purpose or purposes, the more satisfying experiences they can deliver and the stronger their relationships with customers tend to be. In contrast, when companies set out to purchase or build technologies that don't have a clear purpose, the companies tend to get bogged down in technical debt, resulting in a slower delivery of the technology consumers need.

Here are a few specific ways that purpose-driven businesses leverage digital acceleration to advance their mission:

- *Efficiency and Scale.* Digital acceleration allows

purpose-driven companies to deliver products or services more efficiently and to a larger audience. For example, ed-tech platforms like Duolingo and Khan Academy can provide free, quality education to millions of users worldwide thanks to advances in cloud computing and digital content delivery.

- *Innovation.* The rapid pace of technological development encourages continuous innovation, enabling purpose-driven companies to develop new solutions to pressing global challenges. Tesla leverages advancements in battery technology and AI to improve the performance and affordability of its electric vehicles.
- *Data and Insights.* Digital acceleration also means an increase in the amount of data generated and the tools available to analyze it. Purpose-driven companies can use these insights to understand their impact better, make data-driven decisions, and demonstrate their value to stakeholders. Healthcare companies can leverage data analytics and AI to improve patient outcomes and reduce costs.
- *Connectivity and Collaboration.* Digital technologies enable greater connectivity and collaboration, allowing purpose-driven companies to engage meaningfully with their customers, partners, and communities. Social enterprises can use digital platforms and social media to raise awareness, mobilize support, and drive action toward their cause.

We'll have more to say about digital acceleration in the next chapter, but I want to highlight that *purpose-driven digital acceleration* can be a match made in heaven. By embracing digital acceleration purposefully, with foresight and careful planning, businesses become better able to innovate and adapt as needed in pursuing their mission—all at the clip of a rapidly evolving market.

Purpose Drives Customer Transformation

Customer Transformation is closely linked to being purpose-driven because companies with a clear purpose often focus on addressing the needs of their customers and the broader community. By being attentive to the customer experience and aligning with customer values, purpose-driven companies can foster solid, long-lasting relationships with their customers while also turning a profit and making a meaningful impact on the world.

Here are four ways that maintaining a clear, people-centered purpose can help a business advance its customer transformation goals:

1. *Providing solutions for customer needs and challenges.* Purpose-driven companies often develop solutions to help customers overcome challenges or meet their needs in a way that is genuinely beneficial to the customer in addition to being socially, environmentally, or ethically responsible. By focusing on the customers' well-being and the greater good, such companies create products and services that are valuable and meaningful to their customers. An excellent example is CVS Health (formerly CVS Pharmacy), which decided to stop selling tobacco products in 2014. As Carol Cone recounts,

   CVS realized tobacco didn't align with its purpose: "helping people on their path to better health." [But] CVS didn't just remove tobacco from its

stores; it launched several programs to help smokers quit . . . While CVS lost $2 billion in annual cigarette sales in the first year of its new policy, its pharmacy sales jumped . . . These changes resulted in a 10% increase in revenue, notably via growth in pharmacy benefits management.
2. *Building trust and loyalty.* Purpose-driven companies tend to be transparent about their mission and values, which can resonate with customers who share similar values or are passionate about the same issues. One straightforward example pointed out by Max Firsau is The Honest Company, "which was created to deliver 'safe products, simple solutions and clear information' to consumers so they can make better decisions." An alignment of values between the company and its customers can result in greater trust and loyalty, as customers feel that the company genuinely cares about its well-being and the broader impact of its products or services.
3. *Engaging customers and employees in the mission.* Purpose-driven companies often involve their customers in achieving their mission, whether by soliciting product feedback, encouraging customers to contribute to the cause, or fostering a sense of community among users. Such engagement strengthens the relationship between the company and its customers and enhances the business's overall impact. Similarly, it is crucial for a company to promote among its employees "a shared understanding of what the organization is working toward, why it's necessary for business success, and the role each employee plays." Indeed, research

suggests that "[e]mployees are three-and-a-half times more engaged when they understand how their work contributes to the organization achieving its goals."
4. *Fostering long-term relationships.* Purpose-driven companies can foster long-term relationships by focusing on creating a positive impact and aligning with customers' and other stakeholders' values. These companies are more likely to prioritize customer and employee satisfaction and well-being over short-term financial gains, which can lead to greater customer and employee loyalty and retention.

Case Studies

Firsau reminds us, "A purpose-driven business isn't just a trend or strategy deployed to woo customers, but rather, it is part of your brand's DNA which your business lives by." Here are a few examples of companies that have exemplified the benefits of a genuinely customer-centric, purpose-driven approach.

Duolingo

Duolingo is a free language learning platform that has transformed how people learn languages with the use of modern technology and a "mission-driven approach." With a goal to build "the most sophisticated education platform in the world" and make language education available to everyone, "Duolingo works on sophisticated data analytics and artificial intelligence, making it easier for learners to stay motivated, master new material, and achieve their learning goals." The company's gamified, user-friendly experience and extensive language offerings have attracted millions of users worldwide, making Duolingo "the world's most popular way to learn languages and the top-grossing app in the education category."

To put Duolingo's scale in context, there are more people in the United States learning languages on Duolingo than there are foreign language learners in all United States high schools combined, and more people are learning specific languages on Duolingo, like Irish and Hawaiian, than there are native speakers of those languages worldwide.

Duolingo's success is directly tied to its central purpose of making language education as accessible as possible, and its early growth was a direct result of the decision to offer free content. As Natasha Mascarenhas shares:

> Co-founders Luis von Ahn and Severin Hacker never wanted to charge consumers for access to Duolingo content, a purpose imbued throughout the company's culture. For years, to work at Duolingo you had to be comfortable with joining a company in Pittsburgh that was in no rush to make money.

In time, Duolingo was forced to develop a more concrete revenue model to begin turning a profit, but it has continued to prioritize providing a free, high-quality, user-friendly language learning platform to make language education accessible to all:

> Today, Duolingo has a simple freemium business model that is remarkably unconventional. It has a free version with all of its learning content, and it charges a subscription of $6.99 per month for paywalled fea-

tures such as unlimited hearts, no advertisements, and progress tracking. It also has several other revenue streams it's developing, such as language proficiency tests.

Duolingo's purpose-driven use of technology not only benefits individual customers by helping them learn new languages at no cost but also promotes cultural exchange and understanding on a global scale.

Tesla

As the subheading of a Max Pressman's article suggests, "Whether or not you like Elon Musk, he's proven he knows how to establish a business with purpose." One of Musk's most famous ventures, Tesla, was founded to accelerate the world's transition to sustainable energy. The company's innovative electric cars, battery storage systems, and solar energy products have provided customers with environmentally friendly transportation and energy options and spurred the entire automotive industry to invest in electric vehicles.

Tesla's approach balances traditional business purposes (profit-making and expanding market share) with a contemporary purpose-driven approach. On the one hand, Tesla uses conventional business strategies, such as expanding production facilities and entering new markets, to grow and aggressively compete in the automotive industry. On the other hand, the company continually innovates and invests in sustainable energy technologies to drive change in the industry and the widespread adoption of environmentally-friendly solutions. By pushing the boundaries, Tesla has directly impacted its customers and contributed to reducing greenhouse gas emissions on a global scale.

Zipline

Zipline is a cutting-edge electric aerial drone delivery company with a mission "to create the first logistics system that serves all humans equally." Until recently, focused almost exclusively on delivering medical supplies such as blood and vaccines to remote and hard-to-reach areas, Zipline's technology has benefited countries like Rwanda and Ghana, where access to healthcare can be challenging due to difficult terrain and underdeveloped infrastructure. The company is now expanding into various branches of delivery services worldwide, from retail and e-commerce to restaurants to agricultural supplies.

Zipline has already significantly impacted the healthcare system in underdeveloped countries, but its positive impact hasn't stopped there.

In addition to making goods more accessible to people, the company also assists with other regional issues. For instance, using aerial drones means fewer delivery vehicles on the road and, thus, less traffic. Also, using renewable energy reduces air pollution, carbon emissions, and the use of fossil fuels.

Regarding financial impact, Zipline's valuation has steadily increased, reaching $4.2 billion as of an April 2023, proving that a commitment to benefiting humanity is not incompatible with more traditional business purposes.

Purpose: Now More than Ever Before

Purpose can not only be challenging to pin down but can also mean different things to different people. Every organization doesn't perceive purpose as an overarching ideal. Some treat it merely as a marketing tool to project their identity and principles to gain a larger market share. Others contend that offering quality products at the most competitive prices is all that truly matters to consumers. Although there are examples of successful companies that align with such perspectives, the research suggests that longevity and authenticity differentiate purpose-driven businesses from others. Businesses that lead with a defined purpose and structure their operations around it are more likely to earn sustained loyalty, demonstrate consistent performance, and remain relevant in consumers' lives. Conversely, organizations that neglect to recognize and articulate their purpose may manage to thrive in the short term, but consumer expectations will likely increase over time.

Having a clearly defined, people-focused purpose is becoming increasingly important in today's world, and the trend is predicted to intensify. The focus on purpose is mainly due to younger generations who possess a profound sense of purpose, surpassing that of prior generations, and prefer products that tangibly support causes they are passionate about. Furthermore, these younger generations, particularly millennials, are drawn to workplaces with a genuine purpose. Over 70 percent of millennials anticipate their employers to concentrate on societal issues or mission-oriented challenges.

This observation refers back to our treatment of purpose in Chapter 9, where I called purpose the "heart" of the PRAISE method for cultivating a culture of customer transformation in an employee community. In this chapter, I've taken a deeper dive into the centrality of purpose for customer transformation and have argued that a business's purpose must be grounded in *people* if that purpose is to have the staying power to fuel an organization's culture, inspire its customers, and invoke trust, loyalty, and passion among a business's stakeholders.

TWELVE

DATA-DRIVEN RESPONSIBILITY

"In God we trust, all others must bring data."

W. Edwards Deming

I Got that Gut Feeling

How many times have you sat in a boardroom or sales meeting and witnessed a conversation like this between Business, Financial and Technical executives:

BizExec: We need to choose a data-services vendor. I was impressed by the proposal from ABC Data Services.

FinExec: I agree. Their offering was tailored very well to our needs. It also was cost-effective and would certainly produce considerable savings over time.

TechExec: But what about XYZ Corp? I've had good experiences with them.

BizExec: Yes, but from the data we've seen, ABC's solution would offer more benefits and seems to be a better fit for our growth plan.

TechExec: Still, I believe XYZ Corp is a safer bet. They've been around for a long time and have proven reliability.

FinExec: Sure, but ABC's solution is less expensive and seems to be offering a competitive product. Going with XYZ doesn't seem like a sound financial decision just because we're familiar with them.

TechExec: I understand, but I still think XYZ Corp is the right choice. Tech isn't just about costs; it's also about reliability and familiarity.

BizExec: From a business growth standpoint, ABC's product might give us an edge, but I can see your points, too.

TechExec: I think we should go with XYZ Corp. I prefer them.

FinExec: If it's within the budget, I suppose that's fine.

BizExec: I see the value in ABC's solution, but if you prefer XYZ Corp, go ahead with them.

TechExec: It's settled, then. We'll go with XYZ Corp.

It may seem obvious, but how can this kind of decision-making happen–especially in the experienced, professional world? In this scenario, the business and financial executives saw clear advantages to accepting ABC's proposal. However, despite XYZ's technology being more expensive and less able to do what the business needs, the technology executive's "gut" feeling about XYZ company won the day. This may remind you of the story I shared in Chapter 1 of the abandoned shopping carts, where the CEO based a decision solely on a personal belief or bias rather than data.

Confirmation bias is a common type of cognitive bias that impacts decision-making. It typically occurs when a person allows a pre-existing preference or bias to downplay contrary evidence. This sort of thinking can easily lead to skewed conclusions and poor decision-making. Such cognitive bias can occur on an individual basis both in our personal and professional lives, but when it becomes institutionalized in a company (through factors such as internal politics, leadership styles, industry influences, and history), it is known as organizational bias.

In the context of technology leadership, confirmation or organizational bias can draw upon a variety of sources, such as:

- Personal preference for a brand or technology
- Past experiences with a vendor or technology
- Resistance to change or unwillingness to adapt to new technologies

While these biases can sometimes lead to positive outcomes if the favored technology aligns with the company's needs, biased decision-making can also lead to adverse or subpar outcomes if the bias causes the leader to overlook superior solutions. Moreover, when leaders don't take objective data seriously, the effect on subordinates can be frustration, disillusionment, and a toxic corporate culture. Bias decision-making within an organization can undermine customer-centricity processes and destroy customer loyalty.

The Remedy for Bias: Data-Driven Decision Making

Bias is inherently subjective, often fueled by overgeneralizations, jumped-to conclusions, and gut feelings. The best remedy is a healthy dose of objective data. Data-Driven Decision Making (DDDM) is a process that involves collecting data based on measurable goals or Key Performance Indicators (KPIs), analyzing these data, then using the resulting conclusions to make informed decisions. Data and thorough analysis will likely be more objective and consistent than those based solely on intuition or limited scattered observations.

Success stories generated by a careful data-driven decision-making approach are numerous. Here are just a few examples from Tim Stobierski on a Harvard Business School Online blog post to spur your appetite:

1. *Leadership Development at Google* - As part of one of its well-known people analytics initiatives, Project Oxygen, Google mined data from more than 10,000 performance reviews and compared it with employee retention rates. The company then used the information to identify common behaviors of high-performing managers and created training programs to develop these competencies across the company. These efforts boosted median favorability scores for managers from 83 percent to 88 percent.
2. *Real Estate Decisions at Starbucks* - After hundreds

of Starbucks locations were closed in 2008, then-CEO Howard Schultz promised that the company would take a more analytical approach when identifying future store locations. Starbucks now partners with a location-analytics company to create data such as demographics and traffic patterns when determining the likelihood of success before taking on a new investment.

3. *Driving Sales at Amazon* - Data is used to decide which products to recommend to customers. Rather than blindly suggesting a product, Amazon uses data analytics and machine learning to drive its recommendation engine based on prior purchases and search patterns. McKinsey estimated that, in 2017, 35 percent of Amazon's consumer purchases could be tied back to the company's recommendation system.

The Arc of Data-Driven Decision Making

Although different companies approach DDDM in a variety of ways, the overall arc of data-driven decision-making contains the following steps:

1. *Determine what data is needed in light of your business objectives.* Remember that for a customer-centric business, your objectives and the decisions you make to achieve them should always consider your customers' expectations and needs. Some of the most essential data will include the most effective means of interacting with your customers, securing their loyalty, and fostering a strong customer community. Focusing on people (i.e., customers, employees, and other stakeholders) as your prima-

ry purpose (as discussed in the preceding chapter) means you'll need reliable data regarding those key relationships—what makes them tick and what keeps them healthy for your business.
2. *Collect the data.* Organizations collect data from various internal and external sources, including business operations, customer interactions, market trends, and social media. In some cases, the data you need may already have been assembled by some source within your organization or externally. In other cases, you need to consider ways to collect the relevant data independently.
3. *Process and analyze the data.* Once collected, the data must be processed and analyzed to identify patterns, trends, and correlations. You may use advanced analytics tools, statistical methods, and AI in this step, and you'll need to decide on the best ways to visualize the data (e.g., using charts, graphs, and maps) so that it is accessible to all who will participate in the decision-making process. In some cases, you'll need specialists trained in data science to get the most out of your analysis.
4. *Extract insights.* Based on the analysis, you'll seek insights that help your organization understand what the data mean and the implications for decision-making. For instance, the data might reveal that a specific product is selling better in one region or that customers of a certain demographic are more likely to purchase specific services.
5. *Make decisions based on these insights.* With these insights, your organization's leaders can make informed decisions about various aspects of the business, such as which products to push, which

markets to target, or how to optimize operations.
6. *Evaluate the results of your decision-making.* After making decisions and taking steps, you'll want to track and compare the outcomes to the relevant expectations or goals. This allows you to evaluate whether your organization's data-driven decisions were effective and to adjust strategies if needed.

The ultimate goal of DDDM is to bring about desired outcomes by reducing guesswork and subjectivity. DDDM can lead to greater efficiency, more innovation, and a better understanding of market trends and customer preferences, among other benefits. It's a crucial part of modern business strategy, particularly in industries with large amounts of data.

Data-Driven Metrics: OKRs and KPIs

DDDM can be crucial for setting and refining two important tools businesses commonly use when setting targets, measuring progress toward those targets, and continuously improving performance, namely, Objectives and Key Results (OKRs) and Key Performance Indicators (KPIs). Conversely, OKRs and KPIs can provide data for future DDDM. The relationship between DDDM, OKRs, and KPIs is thus complex, and all three together are needed to create a data-informed framework for business strategy and performance management.

In brief, OKRs are "a framework for turning strategic intent into measurable outcomes for an organization." They aim "to connect company, team, and personal goals to measurable results while having all team members and leaders work together in the same, unified direction." DDDM can help set realistic and relevant OKRs. That is, by analyzing historical data and current trends, companies can set achievable objectives and measurable key results.

Once OKRs are defined, companies use KPIs to measure progress toward their objectives. KPIs are specific metrics that provide a quantifiable measure of performance in various areas of an organization's activities. They can be considered a subset of a company's key results and it is vital that they be objective: "As part of defining a KPI, you must determine a projected result for that KPI. If possible, this should be a mathematically-predicted number. It is key to make projections based on actual data. If this is not possible, then periodically agree on—and review—specific targets for each KPI."

DDDM comes into play when interpreting KPI data and deciding on next steps. For example, if a KPI indicates that a company is falling short of a KR, leaders can use additional data to understand why and decide how to act.

Over time, DDDM can also inform adjustments to OKRs and KPIs themselves. If data shows that certain objectives aren't serving the organization well or some KPIs aren't providing useful information, companies can refine OKRs and KPIs accordingly.

Finally, DDDM can also be used to predict future trends and performance. Such predictive analysis can inform the setting of future OKRs and KPIs.

Data-Driven Digital Acceleration

As discussed in the preceding chapter, digital acceleration refers to the rapid adoption and integration of digital technology into all areas of a business. It's a transformative process that changes how companies operate and deliver value to their customers. DDDM plays a crucial role in this process in several ways:

Identifying opportunities. By analyzing internal and external data, businesses can determine where digital technologies

will likely impact a company's ability to fulfill its mission significantly. This could be areas where processes are inefficient, customer needs are not being met, or new technologies could create opportunities for innovation and growth.
Guiding investments. DDDM can also be used in the technology selection and vetting process. By analyzing data on business performance, customer behavior, and market trends, businesses can decide which technologies and associated strategies will likely deliver the best return on investment. As Matt Van Veenendal observes, "It's essential to take enough time to review all data before adopting a digital acceleration strategy. Challenge everything, ask questions, and ensure all is as it should be. Check the technology, the plan, the number of team members, and their knowledge and expertise level."
Driving adoption. DDDM can also assist in driving the adoption of digital technologies within a business. By "using the data-driven approach to analyze the benefits [new technologies] bring to the business," leadership can use data to demonstrate the positive impact of these technologies and thereby help overcome resistance to change and encourage employees to embrace digital tools.
Optimizing performance. Once digital technologies are in place, businesses can use DDDM to optimize performance. This could involve analyzing data to identify workflow bottlenecks, understanding how customers interact with digital products or services, or using AI and machine learning to automate and improve processes. "Collect data and use it to improve the implementation process. Ask employees for feedback through surveys, and monitor data such as software usage statistics and employee productivity metrics to get the most helpful information."

Enhancing customer experience. DDDM enables businesses to understand their customers' needs and behaviors deeply. This understanding can be used to create personalized experiences, recommend products, and proactively address customer needs, which are increasingly important in the digital age.

Mitigating risks. Digital acceleration can bring potential risks, such as cybersecurity threats. DDDM can help identify these risks early and develop strategies to reduce them.

Once the decision to use DDDM to drive digital acceleration has been made, one practical challenge must be how the data will be gathered and analyzed. Ajay Khanna explains why this is no small feat, "the rise of cloud computing and IoT technology, companies have more data at their disposal than ever before," and "many businesses don't have sufficient resources to manage it well." Traditional approaches to analytics may not be up to the task, and to make matters worse, as a Gartner survey reveals, there often aren't enough data specialists:

> Legacy analytics tools are often better at providing reactive rather than proactive insights. On top of that, many companies don't have enough context around their data to make confident decisions or hypotheses. Data tends to give us the "what," not the "why," behind business performance. As a result, it's up to data teams to uncover the whole story, make connections between causes and effects, and recommend changes that improve long-term outcomes. But with data talent in short supply, organizations need more than numbers; they need robust approaches to organize, interpret, structure, and present data meaningfully.

Under such circumstances, automation emerges as an optimal solution for analyzing data on a large scale. It addresses numerous hurdles associated with data analytics, allowing teams to promptly scrutinize all potential data permutations across all existing variables, eliminating the need for analysts to execute singular SQL queries to examine individual hypotheses manually. Furthermore, it prevents analysts from imposing personal biases on their data analysis, enabling them to rely on decision intelligence to study all aspects objectively. Automation of the envisioned sort includes AI, Natural Language Processing (NLP), Machine Learning (ML), and real-time analytics within cloud data warehouses and lakes.

Data-Driven Customer Alignment

Customer-centricity is all about placing the customer at the heart of your business decisions, with the understanding that delivering a superior customer experience will drive business growth. As suggested earlier, DDDM can play a significant role in this approach by providing valuable insights into customer behavior, preferences, and needs. Here are several ways DDDM can align with a customer-centric strategy:

Understanding the customer. Through data analysis, a business can identify key customer segments, gain a more thorough and informed understanding of the customer journey, and discover their needs, preferences, and pain points. These insights can (and should) guide product development, marketing strategies, and customer service improvements.

Personalization. One crucial aspect of improved customer understanding is that DDDM allows businesses to tailor their products, services, and communications to individual customer needs, leading to a more personalized experience.

This can range from recommending products based on past purchases to tailoring marketing messages to specific customer segments.

Customer feedback and satisfaction. Analyzing customer feedback surveys, online reviews, and social media data can help businesses understand their customer satisfaction performance and where they need to improve.

Predictive analysis. DDDM can also help predict future customer behavior and trends. This enables businesses to meet customer needs rather than react proactively.

Continuous improvement. By continuously collecting and analyzing data, businesses can track the effectiveness of different approaches and make changes as needed, constantly refining strategies to serve customers better.

Customer retention. Data can help identify early warning signs of customer churn, enabling companies to take action to improve customer retention.

In short, a data-driven approach to decision-making allows companies to gain a more objective, comprehensive, and nuanced understanding of their customers, with benefits ranging from the ability to better identify potential new customers, on the one end, to an improved ability to retain your customers and foster loyalty on the other.

Data Is King

According to a 2022 New Vantage Partners Data and AI Leadership Executive Survey, although the vast majority of participating organizations (97%) are investing in data initiatives (including 91% who are investing in AI activities), only 26.5% reported that they had succeeded in creating a data-driven organization. Even fewer (19.3%) said they had established a data culture. Obviously, organizations are a long way from DDDM fulfilling its promise to revolutionize how decisions are made in the business world.

Yet, an article on Security Boulevard makes one thing is clear: "In today's fast-paced and data-driven world . . . [y]ou can no longer rely solely on decisions based on your 'gut.'" A rigorous focus on the data has the potential to diminish the effect of subjective biases that might otherwise hinder a genuinely customer-centric approach. That said, the still low success rate of achieving full DDDM implementation suggests that the approach must be thoughtfully applied to have a reasonable chance. Executing a data-driven strategy involves more than just accumulating and examining data. It also necessitates a readiness to evolve and incessantly refine processes as new data and insights emerge. The approach implies pinpointing vital performance metrics, collecting pertinent data, effectively analyzing it, and sharing the results with interested parties. Embracing such a systematic strategy allows for a clear, objective outlook that helps guarantee that the customer's needs and desires remain the focal point of all business choices.

ACTION PLAN

STAGE SIX: TECHNOLOGY

Introduction

The sixth stage concerns understanding the company's purpose and aligning it with customer needs. Without customers, there's no business. You can develop technologies that boost customer relationships and satisfaction by defining a customer-centric purpose and linking it to the company's mission. Companies that fail to consider the customer when creating technologies often end up slow to market, with less customer loyalty and large amounts of technical debt. Furthermore, this stage emphasizes the importance of data-driven decision-making when choosing technological directions instead of relying on personal opinions. Done right, digital acceleration fosters customer loyalty and more efficient market navigation.

Day 0: Reflection and Goal
- **Reflection:** Consider your company's purpose and assess its alignment with your customers' needs. How do your existing technologies embody this alignment, and how effectively does your company utilize data to steer its technological course?

STAGE SIX: ACTION PLAN

- **Goal:** The central goal of this stage is to align the company's purpose with customer needs in such a way that it accelerates digital innovation. We aspire to shape technology around a customer-centric purpose, fostering enhanced customer relationships and satisfaction. Before we embark on this stage, we should maintain a mindset that values alignment and acceleration – acknowledging that your technology decisions need to reflect your customers' needs and that your course should be directed by data, not a personal bias.

Day 1: Content Walkthrough
We'll discuss the significance of aligning company purpose with customer needs and how this alignment can accelerate digital innovation. We'll also delve into the importance of data-driven decision-making in dictating technical direction.

Day 1: Workshop Questions
1. How can aligning the company's purpose with customer needs impact technology development?
2. What role does data play in deciding the direction of our technological advancements?
3. How does digital acceleration influence customer loyalty and market navigation?
4. Can we identify areas where our company's technology doesn't meet customer needs?
5. How can our company reduce technical debt through a deeper customer purpose?

Day 2-7: Homework and Next Steps
Reflect on your company's purpose and how its technology

aligns with this purpose and customer needs. Identify areas where data can better guide your technical direction.

Day 8: Touchpoint and Next Steps
1. Reflect on your company's purpose and alignment with customer needs.
2. Discuss potential strategies for data-driven decision-making in technological development.
3. Identify opportunities to accelerate digital innovation.
4. Develop an action plan to leverage technology in fulfilling your company's purpose.
5. Schedule the next meeting to review and refine this action plan.

Day 9 and Beyond: Action Plan

30-day plan: Begin aligning your customers' goals with your technology purchases, and implement data governance and data-driven decision-making.

60-day plan: Address areas within the organization for digital acceleration, measure the impact on customer satisfaction with that acceleration, and reduction of technical debt.

90-day plan: Evaluate the effectiveness of these changes and their impact on market navigation. Refine your approach based on the results and feedback, and prepare for the final stage of the training.

STAGE SEVEN: BUSINESS

Obsess over Value Alignment

THIRTEEN

THE VALUE OF OBSESSED LEADERSHIP

"A leader is a dealer in hope."

Napoleon Bonaparte

Flag on the Play

  The Great Bud Light War of 2023, dubbed by National Review, erupted into the public sphere on April 1. The tinderbox was ignited by a promotional video from Dylan Mulvaney, a transgender TikTok influencer, promoting Bud Light's NCAA March Madness contest on Instagram. The collaboration resulted in a fierce backlash, with numerous social media users voicing disapproval toward Bud Light for their partnership with a transgender figure. In its wake, a sweeping boycott of the product heavily damaged the pockets of Anheuser-Busch InBev, Bud Light's parent company, and the effects were still being felt four months later.

  Behind the scenes, the alliance with Mulvaney was

merely a fragment of a more extensive campaign initiated by Bud Light's VP of Marketing, Alissa Heinerscheid. The campaign sought to lure a more diverse clientele, targeting primarily the younger demographic and women. Mere days before the video was posted, Heinerscheid had voiced the pressing need to broaden Bud Light's reach: "This brand is in decline; it's been in decline for a really long time, and if we do not attract young drinkers to come and drink this brand there will be no future for Bud Light."

This Bud Light saga serves as a stark case study of inclusive marketing. However, I want to highlight that the core issues extend far beyond that scope. This incident is a profound testament to leadership's inability to maintain a *customer-centric focus during decision-making.*

One glaring failure was the evident oversight by Heinerscheid and the decision-making team to heed the desires and preferences of their established customer base. As management guru Denise Graziano noted in a *Chief Executive* piece, "the biggest error Budweiser made has nothing to do with ideology and everything to do with disregarding their customer base and challenging their sponsorship partners." Her critique goes further: "Bud Light is the official beer of the NFL. Football is arguably the most masculine sport. Ideological opinions aside, it should not have been a surprise that this spokesperson might not resonate with a sizable segment of the NFL audience. A new, more inclusive approach could easily have been market tested in advance, both with message and messenger."

The issue isn't about pandering to certain factions of their customer base. However, as Anson Frericks, a former executive at Anheuser-Busch, accurately summed up, "Bud Light . . . was a brand that was never about politics . . . This is always about a brand that brings people together. It was about football. It was about sports. It was about music. It never got involved in political situations. That's why it was enjoyed by both Republicans and Democrats equally... It was remarkably unpolitical, and this is just a political situation they should not have gotten themselves in." A customer-centric approach would have necessitated careful deliberation before forsaking their historically apolitical stance.

This incident mirrors the second fundamental flaw in Bud Light's leadership: the apparent laissez-faire attitude adopted by the upper echelons of the Bud Light leadership in matters of their brand marketing. A NY Post article explains, "There are reports that senior leadership levels did not know about this campaign. If that is the case, there are failures in management, communication, and chain of command within the organization." In this case, a deeper problem than ignorance is a fundamental disregard for the importance of customer-centricity. Leaders who take their customers seriously must stay informed and involved. Allowing something directly impacting their customer base to slip through the cracks of their management structure shows a clear dereliction of their duty to foster a customer-focused corporate culture.

A Case Study for Leadership Misalignment

Brendan Whitworth, the CEO of Anheuser Busch, responded to the outcry by saying, "We never intended to be part of a discussion that divides people. We are in the business of bringing people together over a beer." This statement, set against the backdrop of foreseeable fallout, lays bare an alarming lack of visionary leadership.

The situation was further exacerbated when, on June 15, 2023, Whitworth, feeling the pressure of plummeting

Bud Light sales, released another message attempting to placate consumers with a simple "we hear you."

Under the banner "Anheuser-Busch Announces Support For Frontline Employees And Wholesaler Partners," Whitworth outlined three critical areas of concentration: "First, we are investing in protecting the jobs of our frontline employees. Second, we provide financial assistance to our independent wholesalers to help them support their employees. Third, to all our valued consumers, we hear you. Our summer advertising launches next week, and you can look forward to Bud Light reinforcing what you've always loved about our brand—that it's easy to drink and enjoy."

Let me ask you: *can you spot the inherent flaw in this list?*

The problem is that the customer—the linchpin of their controversy—is relegated to the bottom of the list. Whitworth offers a placatory, "We hear you," but proceeds to signal a solution embedded in advertising rather than addressing the underlying consumer disquiet.

This hierarchy poses a severe issue as it clouds the real problem. The crux of the matter is not frontline employees or independent wholesalers but rather the disenfranchisement felt by the consumers—the lifeblood of the business.

The list should have commenced with a resounding, "First, and most importantly, we hear our loyal consumers, and we will continue to listen to you." By acknowledging the consumer as the principal consideration behind every decision, Anheuser-Busch leadership would have demonstrated a clear sense of priorities and outlined the appropriate next steps. It's abundantly clear that a mere advertising campaign won't suffice.

Once you rectify the customer problem, the value of

the business and the challenges faced by frontline employees and independent wholesalers will naturally fall into place. Aligning with the customer isn't just a strategy—it's the compass that guides the ship. It should be the leadership's priority, always and unequivocally.

In this context, a quote from Catarina Tucker, founder of Barnastics, becomes particularly poignant: "Consumers are becoming increasingly aware of whether the brands they support align with their values and, as a result, are actively seeking alternatives." The aftermath of the debacle was as swift as it was severe, with Modelo Especial claiming the mantle as "the number one" beer in America just three months after Bud Light's misstep. A throne that Bud Light had safeguarded for 22 long years was lost in a fraction of that time—a testament to how quickly customer loyalty can shift in today's hyper-aware consumer environment.

Customer Obsession

We've talked about customer-centricity throughout this book as a central facet of pursuing a policy of customer transformation. In this chapter on customer-centric leadership, I want to kick it up a notch and, for emphasis, speak about leaders being obsessed with their customers (in a way that Bud Light's leadership appears not to have been).

"Customer obsession" is considered a best practice for businesses aiming to thrive in the age of the customer, where power has shifted from businesses and institutions to customers due to the transparency and ease of access to information brought on by digital innovation. Forrester, a leading global research and advisory firm, has defined customer obsession as "putting the customer at the center of your leadership, strategy, and operations." When a company's leaders constantly and openly return to the questions, "What do our customers truly need and want?" and, "How will this decision or action impact our customers?" it serves as a magnetic force that pulls the organization toward a deep and relentless focus on customer needs and expectations.

Leaders who practice and promote customer obsession understand that it goes far beyond providing good customer service. Customer obsession is about cultivating a deep-rooted culture where every team member is oriented toward understanding and fulfilling customer needs. Leaders inspire this type of culture by setting an example and encouraging a learning environment where customer insights are valued, appreciated, and used for continuous improvement.

This corporate culture of customer obsession, in turn, drives immense business value. Indeed, according to research by Forrester, "investing in customer obsession will yield at *least* a 700% ROI over 12 years, depending on your company and customer type." For one thing, an obsession with customer satisfaction results in high customer loyalty and advocacy levels. Customers become promoters and advocates for the brand, driving referrals and boosting revenue growth. Customer-centric organizations also witness higher customer lifetime value, as satisfied customers tend to make repeat purchases and are willing to try the brand's other products or services.

But the value does not stop there. Internally, organizations prioritizing customer obsession tend to have higher employee satisfaction and engagement levels. This is because a leadership style that prioritizes customers also tends to value its employees. The correlation is direct: When employees feel valued and are part of a purposeful journey, they are more likely to stay, reducing turnover costs and enhancing organizational stability.

Customer obsession is not an overnight transformation but a constant journey that seeks to shape all aspects of the organization—from its mission, vision, strategy, and processes to its success metrics. Because customer obsession is a journey, it must be spearheaded and maintained by visionary leadership who steadfastly take the long view and set the tone for the entire organization. Very little of what I've proposed in this book can be effectively implemented in the absence of this sort of committed leadership. Yet, this kind of leadership will define successful businesses of the future, where value is delivered through consistently excellent customer experiences.

Styles of Leadership for Customer Obsession

It's important to note that there is no one "correct" approach to leadership (more broadly considered) that is required for a customer-obsessed leader. Any leadership style has pros and cons, and any approach can effectively incorporate a healthy obsession with customers that the leadership and its organization serve. Let's consider four common leadership styles: adaptive, purpose-driven, transactional, and transformational.

Adaptive leadership.

This approach, developed by Harvard's Marty Linsky and Ronald Heifetz, "go[es] beyond simply addressing challenges and finding ways to solve them." That is, adaptive leaders are not merely reactive but proactive. They "anticipate challenges and can identify their root causes. Furthermore, they are skilled at recognizing what risks are worth taking and what to avoid wasting the organization's time on." Adaptive leadership, as the name suggests, is especially useful for organizations operating in fluid markets or contexts:

Adaptive leadership helps individuals and organizations adapt and thrive in the face of challenge and prepare them to take on the change process. This leadership approach involves diagnosing, interrupting, and innovating to create capabilities that align with an organization's aspirations.

Regarding customer obsession, adaptive leadership has both pros and cons.

Pros of adaptive leadership:
- Adaptive leaders tend to be flexible and agile without heavy emphasis on rules, allowing them to swiftly respond to changes in customer needs and the broader market.
- Adaptive leaders are open to multiple opinions and encourage a culture of learning and innovation, which can lead to improved customer experience.
- Adaptive leaders expect and embrace change, seeing it as a good thing, thus ensuring their organization remains customer-centric in a rapidly evolving business environment.

Cons of adaptive leadership:
- The relative lack of structure and continual changes can be challenging for some employees to keep up with, potentially affecting service delivery.
- Rapid adaptations may also lead to inconsistent customer experiences if not properly managed.

Effective adaptive leaders will, thus, leverage flexibility and open-mindedness to inspire employees to view changing customer expectations and market fluidity as opportunities rather than threats (see Chapter 10). At the same time, their commitment to customer satisfaction is the one thing that does not change: The mission of customer-obsessed, adaptive leaders is to consistently meet customers' expectations and needs no matter how they may change over time.

Purpose-driven leadership.

Simon Sinek, among others, has championed this leadership style, and, as the label suggests, requires the leader to both perceive the organization's purpose(s) and allow that awareness to guide all strategic thinking, decisions, and actions on behalf of the organization. As an article on All Activity explains,

> Purpose-Driven Leadership is a relational leadership model where leaders put their focus on the core values of their organization and let [those values] guide their actions . . . Purpose-Driven Leadership means disregarding what you believe is important and focusing on what matters most to your customers [emphasis added]. Its philosophy is not about being the best at everything but being the best at the things that matter most. It is all about creating a shared sense of purpose within an organization and aligning the actions of leaders with the greater values and goals. This resonates with the concept of collective leadership, where all organization members work collaboratively toward a shared vision.

Purpose-driven leadership is the style most closely associated with the concept of purpose-driven business discussed earlier in Chapter 11. When applied to customer transformation, the purpose-driven leader will prioritize people, especially the business's customers (as highlighted in the above quotation), identifying what they value with what the business values.

There are, of course, both pros and cons to the purpose-driven leadership style:

Pros of purpose-driven leadership:
- Purpose-driven leaders aim to inspire employees by aligning the organization's mission with a focus on customer satisfaction (hence, customer values).
- Because they passionately focus on what their customers care about, purpose-driven leaders are better positioned to create a strong brand identity that resonates with customers and encourages loyalty.
- The precise sense of attention to customers' values tends to foster consistently high-quality customer experiences.

Cons of purpose-driven leadership:
- If the purpose is too narrowly defined, it may limit innovation and hinder the flexibility needed to meet diverse customer needs.
- There's a risk of employees feeling pressured to uphold the purpose at all costs, which may lead to burnout or neglect of other critical business areas.

Purpose-driven leadership can be highly motivating for an organization's employees and customers alike because it taps into one of the most fundamental needs we as human beings have: to find meaning and significance in our beliefs, commitments, relationships, decisions, and actions. Because the purpose is such a potentially powerful motivating force, the purpose-driven leader must keep their finger on the pulse of employee and customer attitudes to maintain balance and ensure that both groups' genuine needs are being met.

Transactional leadership. This traditional leadership style has a long history and was first described in 1947 by German sociologist Max Weber, then further expounded and contrasted to the transformational leadership style (see below), first by James Burns and additionally by Bernard Bass. Transactional leaders "rely on rewards and punishments to achieve optimal job performance from their subordinates." This methodology is grounded in theories suggesting that individuals lack inherent motivation and require structure, guidance, and supervision to fulfill their job responsibilities. The theory further hypothesizes that employees will carry out their tasks in the manner desired by the transactional leader in return for something the employees desire, such as compensation. Because transactional leaders tend to favor structure and efficiency and are often more reactive and less open to change, a transactional leadership style usually "works best in a structured environment where there are few deviations from established business processes and defined roles with specific tasks to accomplish." For this reason, although other more modern styles are preferred in today's corporate culture, "transactional leaders remain valued in organizations such as the military and large companies where rules and regulations dominate."

Regarding customer transformation, transactional leaders rely on their ability to create appropriate transactions (i.e., rewards and punishments) to motivate employees to be consistently customer-centric. This approach comes with both potential pros and cons:

Pros of transactional leadership:
- In some situations, thoughtfully chosen rewards and punishments can be highly effective in motivating employees to reach specific customer service targets.
- A transactional leadership style can ensure efficiency and consistency, which are essential for maintaining customer service standards.

Cons of transactional leadership:
- A transactional approach may stifle creativity and limit initiative, making it rare for employees to exceed customer expectations and provide exceptional experiences. "Just enough" can quickly supplant "above and beyond."
- Transactional leadership can lead to employee disengagement, indirectly impacting customer satisfaction.

In short, exercising caution is vital when adopting a transactional leadership approach—transactions alone may not prove to be enough to keep employees highly motivated to deliver exceptional customer experiences on a consistent basis. In many cases, the transactional approach may better serve as a supplement to other styles of leadership rather than function as the leader's primary style.

Transformational leadership.
As alluded to above, leadership expert James Burns was the first to speak of transformational leadership in contrast to transactional leadership. As a University of Massachusetts article explains, "transformational leaders know how to encourage, inspire and motivate employees to perform

in ways that create meaningful change." Whereas transactional leadership "focuses more on extrinsic motivation for the performance of specific job tasks," transformational leadership "inspires employees to strive beyond required expectations to work toward a shared vision." To this end, transformational leaders "are not afraid to challenge employees" but at the same time "listen to employees' concerns and needs so they can provide adequate support." An important element of this support is a transformational leader's ability to create a safe, empowered environment for employees. (Recall from Chapter 9 the pivotal role of psychological safety in building a culture of customer transformation.) Transformational leaders foster an environment where employees feel free "to have conversations, be creative, and voice diverse perspectives. This empowers employees to ask questions, practice greater autonomy, and ultimately determine more effective ways to execute their tasks."

The impact of such an environment on customer transformation can be significant. As the style places a heavy emphasis on leading through inspiration, Kendra Cherry defines transformational leaders as "generally energetic, enthusiastic, and passionate" about the mission of the business and its potential impact on customers. "These leaders have a marked passion for the work and an ability to make the rest of the group feel recharged and energized." Such passion and enthusiasm inevitably tend to rub off on employees, who similarly affect the customers with whom they interact. Not surprisingly, a transformational approach can be a good fit for a customer-obsessed leader intent on producing a customer-obsessed corporate culture.

Pros of transformational leadership:
- Transformational leaders inspire and motivate employees to go beyond their assigned tasks, fostering a culture of exceptional customer service.
- Such leaders also promote high engagement and job satisfaction, which tends to translate directly into positive customer experiences.

Cons of transformational leadership:
- If not carefully managed, high expectations can lead to burnout for some employees, affecting employee well-being and customer service.
- There's a risk of overlooking the small details and practicalities of delivering consistent customer service in pursuing wide scale transformational change.

To sum up, transformational leaders are generally well-suited to foster a customer-centric culture because not only can they inspire employees to put customers first, but they also do so within the context of a highly supportive relationship with these same employees. Such support decreases the likelihood that employees will lose focus or burn out and increases the likelihood of success in the company's ongoing customer transformation.

Best Practices for Customer Obsessed Leadership

Regardless of which leadership style or combination of types you feel most comfortable emulating (including other styles not mentioned above), here are 11 best practices that will facilitate your commitment to being a customer-obsessed leader:

1. *Lead by example:* CEOs should embody the customer-obsessed culture they want to cultivate. This could include personally engaging with customers, resolving high-level customer complaints, or regularly discussing customer feedback in company meetings.
2. *Check your ego:* I've been in several board meetings where no one at the table would speak unless the top executive spoke first. This is the sort of hesitancy that arises in the presence of a sizable, unrestrained ego. Check your ego at the door, allowing others to talk while you listen. Chime in when necessary, but keep in mind that you'll never get genuine insights and feedback if the culture is steeped in fear of speaking up in front of the leader. Stop with the "Yes, Boss" syndrome, whether the "boss" in view is a lower-level team leader or the CEO of the entire corporation.
3. *Set clear expectations*: Clearly communicate the vision of a customer-centric organization to all employees. Make sure each team knows its role in achieving this goal.
4. *Develop a customer-centric strategy*: Ensure customer obsession is ingrained into the company strategy. All business objectives and initiatives should align to improve the customer experience.
5. *Prioritize customer experience in decision-making*: Consider the impact on customers in all decision-making processes. This can help avoid decisions that might negatively impact customer satisfaction.
6. *Empower employees*: Allow employees the autonomy to make decisions that can enhance customer

satisfaction. Empowered employees are often more engaged and responsive and can solve customer issues more efficiently.
7. *Foster open communication*: Encourage employees to openly share customer feedback and ideas for improvement with their teams and supervisors. Regularly communicate progress toward customer-centric goals to keep everyone aligned.
8. *Listen more; talk less:* Leaders should spend less time talking (mostly about themselves and their ideas) and more time listening to their employees, stakeholders, and customers. The CEO should periodically make time to interact with actual customers. It would be awesome to see a CEO who holds town hall meetings for their customers or a leader who participates in focus groups or customer advisory boards. Such direct involvement will generate loyalty among employees and customers alike.
9. *Leverage data*: Use customer data to refine your business's understanding of what your customers truly need and want and to drive strategy (see Chapter 12). Consider developing data-driven personas to understand and serve customers better.
10. *Promote continuous learning*: Provide training and resources for employees to continuously improve their customer service skills. This may include formal training options (e.g., e-learning courses, workshops or conferences, mobile learning courses) and informal (e.g., discussion and collaboration on social media, coaching, and mentoring options, relevant blogs).
11. *Celebrate success*: Acknowledge and reward em-

ployees and teams who deliver superior customer service. Doing so not only provides additional motivation to employees, but it also reinforces the paramount importance of the customer to your business.

The Power of Leadership

If it's not already obvious from what I've covered above, let me be clear: There is no one "style" for customer-obsessed leadership. Instead, various approaches to leadership, each with its distinctive strengths, can be infused with a passionate focus on the customer, thereby generating greater value for both the business and the customers. In closing this chapter, I'd like to emphasize that the responsibility for making this happen rests squarely on the shoulders of the leadership. Leaders who make bad decisions that don't align with customers will lose value (= Bud Light); leaders obsessed with knowing and providing for their customers will, conversely, increase value. Few things in life are as straightforward as that.

This power of leadership initiative is illustrated by the story of Mark Cuban's successful efforts to revamp the culture of the Dallas Mavericks professional basketball team. As Utathya Ghosh recounts, "The Mavericks faced a significant challenge when an investigation uncovered evidence of s*xual harassment and workplace misconduct within the organization, predating Cuban's ownership." Exhibiting a "willingness to acknowledge the issues" and a "dedication to creating a safe and inclusive workplace . . . Cuban's leadership style played a vital role in fostering a culture of transformation within the Dallas Mavericks." Cuban handpicked a new CEO, Cynt Marshall, who similarly possessed a "commitment to gender and racial diversity within the leadership team." As a result of their combined efforts, the organization "witnessed a notable shift," in time "boasting equal representation of women and people of color in executive positions" and winning the NBA's prestigious 2022 Inclusion Leadership Award—and, in the course of it all, becoming one of the league's most valuable teams (worth $2.4 billion in 2022). Faced with a grave threat to the reputation of the organization, Cuban had thus correctly perceived the value that an inclusive, representative leadership team would hold for both the organization's members and its customers, and he had doggedly pursued this goal until it was achieved, creating tremendous business value in the process.

In the final chapter, we'll explore in more detail the notion of "value alignment," or the process of making sure that the values embraced by a company's leadership align with customers' values.

FOURTEEN

CUSTOMER VALUE ALIGNMENT

"The purpose of a business is to get and keep a customer."

Theodore Levitt, 1986

The One-Dollar Difference

I've argued throughout this book that businesses that choose to embark on a process of customer transformation—that is, the wholesale recentering of strategies, decision-making, and operations around the customer, seeking to understand, meet, and exceed the customer's expectations in ways that build long-term relationships—will create enduring business value.

But what *is* business value, exactly? While it is a multi-faceted concept \more easily intuited than analyzed, the most prominent mistake executives and salespeople make regarding business value is the failure to appreciate the concept's foundation rather than its complexity. The reference point for business value is *customer* value, or that which matters to your customers. And what your customers value cannot be reduced merely to dollar figures, as the following story illustrates.

During my time at Google, I worked with a team called Value Advisory whose core purpose was to develop value propositions to help the sales teams close deals. The problem we faced, however, was that everything sales *wanted* and value leads *focused on* was reduced to Business Case development—simply quantifying the benefits of Google Cloud's products in financial terms with ROIs and TCOs. Although there was a category of value recognition behind the numbers, this refusal to focus on other value types often led to the loss of substantial deals.

The argument I made to our sales teams about our customers was simple: If our cloud customers can't see the value of what we are offering as it relates *directly to their customers*, instead of to their budgets, it doesn't matter how much they save or what our services cost, they don't and probably won't get it.

The validity of this argument is neatly demonstrated in two separate engagements.

During one meeting with a customer's executive team, we discussed the cost and benefits of a Cloud service. This particular customer, an international entertainment company, used a competitor's product for this service. Our sales team repeatedly asked the customer to reveal our competitor's prices, but the customer wouldn't.

I was asked to be on the call to highlight the value of our service. However, as usual, the client only wanted to talk about money, the sales team was happy to oblige, and no one was interested in the impact the products would have on the client's customers. When the sales lead asked me for help during the call, I calmly said, "To simplify this, let's guarantee that we will beat our competitor's price by $1.00."

Everyone was stunned, and no one said a word for a moment. But then I could see the "ah-ha" moment take hold around the table. It had become clear that the client was using us to get our prices lower to help them negotiate a better price with their existing vendor. The client didn't care about the impact our services would have for their customers. It wasn't about the customers at all.

In a meeting with a different client, I sat around a table with our sales team and their executive team. We had spent 30 minutes discussing a cloud product they were interested in, and I focused heavily on the value for their customers. Toward the end of the meeting, the CTO turned to me and asked, "This all sounds great, but it comes with a cost. What's the price?"

Without skipping a beat, I said, "Does it matter?" The CTO didn't respond, but he looked at me for more information, so I continued. "Suppose I say it will cost you $1 million [far more than the product's actual cost], but that investment will help you gain 100,000 new customers from the improved services and the new products you'll be able to create."

The CTO looked at me, deep in thought, then glanced around at his team and told them, "Let's make this happen."

This conversation didn't require a table of complex numbers to paint a best-case scenario. The leader understood his customers' needs, our product's value proposition, and how his team would be able to differentiate themselves by incorporating our technology. He understood what it would mean to his customers and how that translated into business value.

Value-Based Selling and Customer Value

The technique I employed with the aforementioned CTO is called *value-based selling*, and its success depends wholly on the assumption that what customers value—what they *truly* want—goes far beyond mere dollars and cents. Lestraundra Alfred describes value-based selling as "a philosophy purely rooted in 'solving for the customer,'" where the customer's problem is, essentially, *what do I, the customer, need, and what is the best way to meet that need?* There are four basic categories of customer values:

1. *Qualitative value.* This type of customer value is addressed by the abstract benefits your product or service provides that can smooth out the "hitches, hiccups, and inefficiencies" your customer constantly faces in life or business, that is, their "day-to-day pain points." These benefits can range from something as simple as "ease of use" to something as profound as "personal success."
2. *Financial value.* No one is saying that dollars and cents don't matter—of course they do. It matters to an end user if your product costs less than a competitor's as well as if your offering can generate more revenue and/or lead to lower operating costs than the competition.
3. *Differentiation value.* Business customers need to stand out from their competition, and they'll value your ability to make this happen. "If you can show that your product or service aligns well with whom your buyer wants to be within their competitive landscape, you'll set yourself up for a successful value-selling effort."
4. *Security value.* This value comes from your ability

to address "any specific fears or stressful vulnerabilities" your customer is dealing with. Business leaders face risks ranging from "direct security threats to emerging industry trends they might struggle to keep pace with." Security also touches on customer empathy where users have their own seemingly endless list of sources of anxiety, whether that be threats to their health, a need for more information about an issue, or fear of social or relational rejection.

The point here is that customers care about a wide range of things, any portion of which could be crucial to not only be able to "understand" your customers but also (perhaps even more importantly) how much your product or service will be worth to them (i.e., your product or service's customer value). This broad spectrum of your customers' concerns and objects of value should be your reference point—your North Star—for determining how your business will generate value for its customers.

The Pillars of Business Value

With the above understanding of the nature of customer value, we're now ready to consider the equally multifaceted nature of business value. In broad terms, business value refers to "the entire value of the business; the sum total of all tangible and intangible elements. Examples of tangible elements include monetary assets, stockholder equity, fixtures, and utility. Intangible elements include brand recognition, goodwill, public benefit, and trademarks." The concept of business value can be approached through a cluster of key pillars or broad value types. Here are a few of the more widely recognized ones, with some brief comments about how a commitment to customer-centricity can strengthen each of these pillars:

Customer value: This perspective looks at how well a business meets the needs and expectations of its customers—needs and expectations which, as we just saw, come in many forms. Key elements of the value a business holds for its customers often include product quality, customer service, price, and brand reputation. As we've seen repeatedly throughout this book, high customer value generates numerous rewards for an organization—for example, an increase in repeat business and referrals.

Innovation value: In a rapidly changing business environment, a company's ability to innovate and adapt is a crucial determinant of its long-term value. Through continuous innovation as well as investing in intuitive, user-friendly interfaces, a company can enhance the user experience, making its products or services more appealing and competitive. This can, in turn, lead to increased market share and customer retention.

Strategic value: When I mention 'strategic value', I am talking about fundamental guiding principles that shape a company's decisions, actions, and behaviors. In other words, what the company itself values most as a business. When a business places a premium value on customer satisfaction and an optimized customer journey (i.e., strategically valuing the customer), the company will benefit from stronger customer relationships, an enhanced reputation, and improved consumer lifetime value. These efforts can also result in positive word-of-mouth, leading to new customer acquisition.

Community value: Leveraging digital ecosystems to develop an online community can help an organization extend its reach, improve customer engagement, and generate new revenue streams. Furthermore, fostering a vibrant online customer community can enhance brand loyalty and customer retention and provide valuable insights for product development and innovation.

Cultural value: A business is only as strong as its people. Employee engagement, job satisfaction, professional development opportunities, and workplace culture are vital elements of business value. Undertaking a customer transformation entails a cultural transformation and aligning the organization's customer-centric values and behaviors with its strategic goals, leading to increased employee engagement, productivity, and innovation. A positive, customer-focused organizational culture can also enhance the company's reputation, aiding in talent acquisition and retention.

Acceleration value: This involves how efficiently a business utilizes its resources. The answer will impact all other business value aspects, from financial performance to customer and employee satisfaction. Embracing a customer-conscious digital acceleration strategy (see Chapter 11), in particular, can help a company improve operational efficiency, reduce costs, and quickly respond to market changes and evolving customer preferences. Doing so also opens up opportunities for innovation, new business models, and improved customer experiences.

Financial value: Financial metrics, including profitability, revenue growth, cash flow, and ROI, are among the most direct ways of assessing a business's value. Streamlining business operations can lead to cost savings, improved quality, and better customer service. Combined with prudent financial management, this can increase profitability, providing resources for growth and delivering a higher ROI for stakeholders.

The main point is, by marshaling all available energy and resources to plan for, create, and sustain experiences that meet and exceed customer expectations, the process of customer transformation can potentially increase value in each of the above areas. At the risk of oversimplifying (but it's a risk I'm willing to take), much of this reduces to the following equation:

$$\text{Happy Customers} = \text{Business Value}$$

Measuring Business Value

Measurement plays a pivotal role as we navigate the landscape of customer transformation and business value. Periodic measurements of business value allow companies to evaluate the effectiveness of their customer-centric strategies and better understand how these strategies influence business value.

One crucial set of metrics revolves around customer satisfaction and loyalty, such as Net Promoter Score (NPS), customer satisfaction score (CSAT), and customer lifetime value (CLTV). These metrics can help businesses understand whether they are meeting or exceeding customer expectations and how customer relationships impact their bottom line.

Churn rate and customer retention rate are also essential indicators. They measure the ability of the business to retain customers over time, which is a critical aspect of customer transformation. High retention rates generally indicate that customers perceive high value in the company's offerings and are more likely to remain loyal.

Moreover, businesses should assess the impact of customer-centric strategies on their financial performance. Metrics like ROI and sales growth can provide insights into how customer transformation drives revenue and profitability.

Tracking these metrics over time allows businesses to better understand the correlation between customer transformation and business value. This, in turn, will enable business leaders to make more informed decisions, optimize their strategies, and maximize the business value derived from their customer-centric approach.

Value Alignment

In addition to periodically measuring business value during customer transformation, it is also crucial to ensure that (as much as possible) company values are consistently possessed and expressed by all relevant stakeholders across all spheres of the business's activity and influence. As mentioned at the end of the preceding chapter, the responsibility for such alignment of corporate values, in the words of one observer, "always starts at the top . . . with the CEO and senior leadership." The term value alignment often refers specifically to the congruence between the core beliefs and principles of an organization and the values, behaviors, and attitudes of its employees. Value alignment can also refer to the correlation between the business's values, strategies, operations, and practices. For instance, a company that values sustainability should align its business practices to reflect this, such as implementing recycling programs, reducing energy usage, or sourcing from sustainable suppliers. Even more relevant for present purposes, a company that claims to put its customers first should exhibit this value in its business practices by offering products and services that meet real customer needs, providing exemplary customer service that goes above and beyond customer expectations, and by generally creating exceptional customer experiences as part of a meaningful customer journey.

Here are some essential aspects of value alignment to consider:

Customer alignment. This refers to aligning a company's values with its customers' needs, wants, and values. A company whose values are aligned with those of its customers will be better able to meet their expectations and build strong, long-lasting relationships. If a company's customers place a high value on sustainability, it would be beneficial to incorporate sustainable practices into its operations and communicate these efforts to customers. As discussed in Chapter 12, understanding and aligning with customer values can drive customer loyalty and influence purchasing decisions.

Employee alignment. It's important that employees not only understand but also share the company's core values. As discussed in Chapter 9, such alignment can improve job satisfaction, commitment, productivity, and teamwork.

Leadership alignment. Leaders should exemplify the company's values as they set the tone for the entire organization (see Chapter 13). An inconsistency between leaders' actions and the company's stated values can quickly lead to cynicism and mistrust among employees.

Operational alignment. The organization's operations, business strategies, and daily practices should reflect its values. A company that values customer service should invest in relevant employee training and establish procedures to ensure high-quality service.

External alignment. This is about ensuring that the company's external image, branding, and public-facing communications align with its internal values—including how it interacts with its customers, suppliers, and the wider community. If a company claims to value ethical sourcing, it should be transparent about its supply chain practices.

Otherwise, a disconnect between a company's internal values and its external image can harm its reputation and lead to stakeholder mistrust.

I mentioned the importance of periodically measuring a company's business value, but what about its value alignment? That must be measured as well.

Measuring Value Alignment: Customer Transformation and Value Alignment (CTVA)

In Chapter 13 I emphasized the importance of the flow of values proceeding in a particular way in a customer-obsessed company: The values of the *customers* should be the reference point that inspires and informs the values of the company's *leaders*, which should, in turn, then inspire and inform the values of *employees* at all levels. Thus, value alignment must always begin with considering the customer, and it is up to the leadership to ensure that everyone in the organization is aligned with the values developed on this basis.

This alignment of values must not be taken for granted initially or over time. Periodic assessment is essential:

> When something is assessed, it is addressed. Set up ways to test whether you are acting according to your values. Ask your customers. Hire consultants. Listen to your employees. Get feedback about how aligned your vision and practices are. Even if you get positive results, keep assessing so you do not slip into contrary practices. Staying aligned with your company's core purpose and values can help companies excel.

The *Customer Transformation and Value Alignment* tool (CTVA) presented below offers a multidimensional approach to measuring an organization's alignment with its customer values. Encompassing seven dimensions corresponding to the seven overall stages of customer transformation outlined in this book, the CTVA allows for the assessment of each dimension on a -5 to +5 "Goldilocks" scale, with the numbers representing extremes (too little or too much) and 0 representing the "just right" balance most likely to meet customer expectations. The assessments themselves are sourced both from the company's customers as well as its other stakeholders (i.e., shareholders, various executives, employees, and partners) based on each pool of respondents' answers to a Customer Question (in the case of customers) and an Internal Question (in the case of other stakeholders) for each of the seven dimensions.

Value Alignment Scale

Not Enough					Just Right				Too Much
-5	-4	-3	-2	-1	0	1	2	3	4	5

Value Alignment Metric (Ideal Alignment)

Customer	S.H.	CxO	VP	Dir.	Mgr.	Emp.	Prtn.
0	0	0	0	0	0	0	0

The CTVA is presented in a standard form here but can be customized to your company's specific needs (see the Action Plan following this chapter for a CTVA template along with suggested guidelines for implementation). Let's briefly dive into the seven dimensions of customer transformation measured with the CTVA in its "standard" form:

Stage One: Customer

Customer Question: How well does the company understand and address your needs and preferences?

Internal Question: To what extent do you feel the company understands and addresses customer needs and preferences in your role?

- -5: The company seems oblivious to customer needs and preferences. Customers feel unheard and overlooked, leading to dissatisfaction and missed opportunities for the business. Employees feel the company fails to capture and respond to customer needs and preferences.

- 0: The company accurately understands and addresses customer needs and preferences. Customers feel heard and appreciated. Employees feel that their role contributes meaningfully to understanding and meeting customer needs.

- +5: The company excessively focuses on customer needs and preferences, so it might appear overly eager or desperate. This could lead to customers feeling intruded upon or overwhelmed. Employees might feel the company is excessively customer-oriented, leaving little room for internal improvements or innovation.

Stage Two: Interfaces

Customer Question: How would you rate the company's ability to provide new features or ways of connecting with its products and services?

Internal Question: How well does the company innovate and implement emerging interfaces to enhance the customer experience in your department?

- -5: The company's ability to innovate in order to enhance customer experience is non-existent or highly outdated. Customers struggle with the interface, leading to frustration. Employees find that the company's technology hinders rather than enhances products and services.

- 0: The company's innovation appropriately enhances the customer experience. Customers find the interface user-friendly and value-adding. Employees are often allowed to innovate new ways of connecting products and services with their customers.

- +5: The company innovates excessively, often causing customer confusion or discomfort. While some may appreciate the tech-forward approach, others might find it intimidating or unnecessary. Employees may feel overwhelmed by the constant introduction of new technology.

Stage Three: Journeys

Customer Question: From your first contact with this company until now, how well has the company provided a problem-free, personalized experience during your interactions with them?

Internal Question: How well does the company optimize the customer journey and interconnect with external journeys to create a personalized experience?

- -5: The company provides a disjointed and impersonal customer journey. Customers feel disengaged and unimportant. Employees feel they lack the resources or strategies to offer a seamless and personalized customer experience.

- 0: The company provides a seamless and personalized customer journey. Customers feel valued and engaged every step of the way. Employees believe they can effectively optimize the customer journey.

- +5: The company excessively personalizes the customer journey, which can come off as overbearing or insincere. Some customers might feel overwhelmed by the level of personalization. Employees may feel like they are excessively micromanaging the customer experience.

Stage Four: Ecosystem

Customer Question: How would you rate your interactions with the company's employees, specifically regarding communication, customer service, and idea sharing?

Internal Question: How well do you feel the company's employees listen to our customers to create opportunities to innovate new products and services?

- -5: Interactions with the company's employees are poor, including inadequate customer service and communication. Customers feel undervalued and frustrated. Employees feel they are not equipped to provide exceptional service and communication.

- 0: Interactions with the company's employees are positive, marked by exceptional customer service and sharing of ideas. Customers feel respected and heard. Employees feel that they effectively deliver top-notch service that the community is asking for.

- +5: Interactions with the company's employees are excessively friendly or communicative, potentially making customers feel uncomfortable or overwhelmed. While some might appreciate the high-touch approach, others may find it intrusive. Employees may feel they are pushing too hard to connect with customers.

Stage Five: Culture

Customer Question: To what degree do you feel the company's employees care about your needs and preferences as a customer?

Internal Question: To what extent do you believe the company's culture and values align with a customer-centric mindset?

- -5: The company's culture is not customer-centric, and employees' actions and behaviors reflect this. Customers feel this lack of focus on their needs. Employees think the culture does not support a customer-centric mindset.

- 0: The company's culture is customer-centric, and employees' actions and behaviors align with this. Customers feel cared for and valued. Employees believe the culture and values support a customer-centric approach.

- +5: The company's culture is excessively customer-centric, possibly neglecting employees' needs and business sustainability. Customers might feel the company is too eager to please. Employees may feel that the hyper-focus on customers overlooks their well-being or other business imperatives.

Stage Six: Technology

Customer Question: How well does the company use technology to provide convenient and fast services that meet your needs?

Internal Question: How effectively does the company adopt and utilize technology to deliver efficient and customer-centric services?

- -5: The company does not effectively leverage technology, leading to inconvenient and slow services. Customers are frustrated by the lack of modern solutions. Employees feel hindered by outdated or ineffective technology that doesn't have purpose.

- 0: The company effectively leverages technology to provide convenient and fast services. Customers appreciate efficient, tech-enabled solutions. Employees feel empowered by the technology to deliver efficient and customer-centric services that have a purpose meaningful to them.

- +5: The company excessively leverages technology, possibly neglecting the human aspect of service or overwhelming customers with complex solutions. While some customers might appreciate the advanced technology, others might find it confusing or impersonal. Employees may feel they're overly reliant on technology.

Stage Seven: Business

Customer Question: How well do the company's products and services and the way the company operates match what you seek as a customer?

Internal Question: How well do the company's offerings and operations align with the value customers seek in your department?

- -5: The company's offerings and operations do not align with the value customers seek, leading to dissatisfaction and lost opportunities. Employees feel the company is misaligned with customer needs and values.

- 0: The company's offerings and operations align well with the value customers seek. Customers feel the business understands and caters to their needs. Employees believe the company's operations and offerings meet customer expectations and values.

- +5: The company excessively tries to align its offerings and operations with perceived customer value, possibly coming off as desperate or pushy. Customers might feel overwhelmed by the intensity of the company's efforts. Employees might feel the company is trying too hard to please customers at the cost of other vital factors.

Although this version of the CTVA is based on the seven dimensions of customer transformation covered in this book, the CTVA can be modified and used to assess any area of your organization regarding the degree of alignment between the values of your customers and the various stakeholders in your company. Refer to the template in the Action Plan after this chapter for more information.

Benefits of the CTVA

Focusing on the few at the expense of the many can lead to dissatisfaction for most. By helping to ensure that the values held by a customer-centric company are thoroughly aligned, the use of the CTVA will benefit companies in several important ways:

First, companies can strengthen customer satisfaction and loyalty by verifying the alignment of all stakeholder values with customers values. This fosters repeat business and positive word-of-mouth within a growing customer community.

Second, the CTVA provides valuable insights that can inform organizational decisions and strategies. This can impact everything from product development and pricing to technology adoption and leadership initiatives.

The CTVA also promotes employee engagement. By aligning employee actions with customer values and showing employees how their roles contribute to customer satisfaction, employees tend to be more motivated to help create exceptional customer experiences.

By demonstrating their commitment to customer values, companies enhance brand reputation and differentiate themselves from competitors. This tends to increase long-term customer satisfaction and loyalty.

Lastly, the ability to measurably track and ensure alignment with customer values over time makes businesses more agile and responsive to changes in customer expectations or market conditions.

I encourage you to take some time with the following Action Plan and consider how you might use the CTVA to assess the state of value alignment in your own company. In my consulting career, I've witnessed too many business leaders default to a head-in-the-sand approach, initially dismissing the need for an objective assessment because the executive "already has a sense of things" in their own company. I've also seen more than a few such leaders who were surprised or even shocked by what they found when they reversed course and carried out an assessment like the CTVA. There is no substitute for building periodic, objective reviews into your long-term business strategy.

ACTION PLAN

STAGE SEVEN: BUSINESS

Introduction
The seventh and final stage brings us to operations, applying all the work from the previous six stages to generate business value. This stage zeroes in on leadership, metrics, and understanding the actual value for a business: your customer. This training aims to improve the alignment between your customers and your organization. Happy Customers = Business Value. We will examine Customer Value and Transformation Alignment and integrate the process into your long-term business strategy.

Day 0: Reflection and Goal
- **Reflection:** Contemplate how your business currently perceives and defines value. Reflect on how your leadership aligns with the principle that the customer is the core purpose of your business. Assess the metrics currently in place to measure the alignment of your business value with your customers' values.

- **Goal:** The primary objective of this final stage is to operationalize all the insights and transformations from the previous stages to generate tangible business value. The aspiration is to align your customers more closely with your organization, with an understanding that satisfied customers equate to real business value. We aim to scrutinize Customer Value and Transformation Alignment and incorporate this process into your business strategy. As we begin this final stage, we should embrace a mindset of purpose-driven leadership, understanding that your success lies in your obsession with delivering to your customers' aspirations.

Day 1: Content Walkthrough
We'll explore the relationship between customer satisfaction and business value. We will also delve into leadership, different types of value, and the Customer Value and Transformation Alignment metrics to track your progress.

Day 1: Workshop Questions
1. Considering the customer's primary purpose, how can our leadership approach be more purpose-driven or obsessed?
2. How does customer satisfaction currently influence business value at our company?
3. What forms of value does our business recognize, and how could this be expanded?
4. What role do KPIs and OKRs play in tracking the value's progress?
5. How can Customer Value and Transformation Alignment be implemented in our business operations?

Day 2-7: Homework and Next Steps
Reflect on your business's definition of value and how it's tied to customer satisfaction. Identify potential KPIs or OKRs that could better track your progress.

Day 8: Touchpoint and Next Steps
1. Share your reflections on business value and its relationship with customer satisfaction.
2. Discuss potential strategies for ensuring your leaders are obsessed with your customers.
3. Implement your first Customer Value and Transformation Alignment campaign.
4. Develop an action plan to increase business value through customer value alignment.
5. Schedule the next meeting to review and finalize this action plan.

Day 9 and Beyond: Action Plan
30-day plan: Begin implementing leadership that obsesses over your customers and redefines business value with a focus on customer satisfaction.

60-day plan: Implement Customer Value and Transformation Alignment, and integrate the process with Product Management and Customer Service teams.

90-day plan: Evaluate the effectiveness of these changes, their impact on business value, and customer satisfaction. Refine your approach based on the results and feedback, and celebrate the completion of the training program.

Ready for the next step?

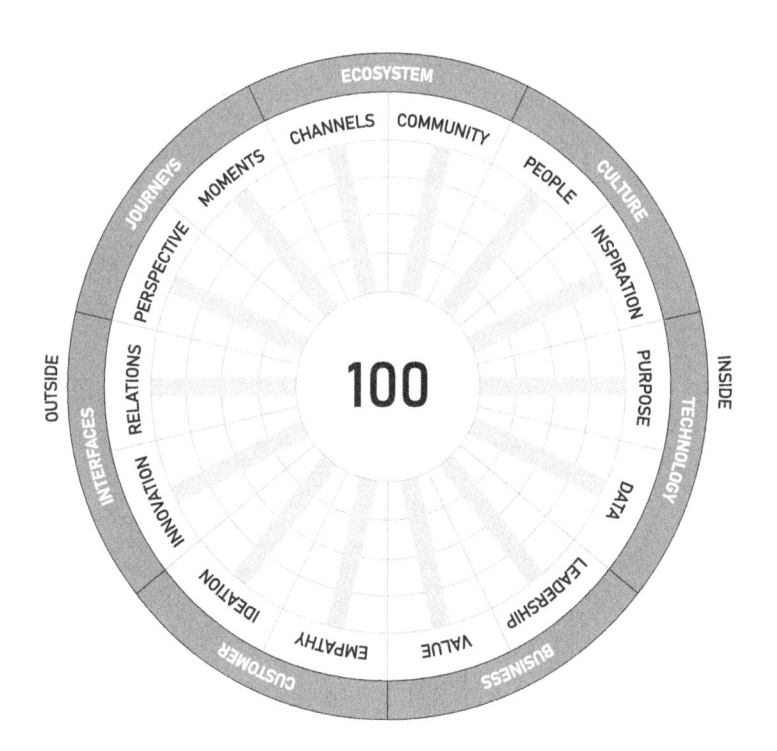

CUSTOMERTRANSFORMATION.com

Take the free assessment to learn where your organization stands across the seven stages of customer transformation.

ENDNOTES

Chapter One

Zavery, A. (2020, January 22). "*Digital Transformation Isn't a Project, It's a Way of Operating.*" https://www.forbes.com/sites/googlecloud/2020/01/22/digital-transformation-isnt-a-project-its-a-way-of-operating/

Vial, G. (2019). Understanding digital transformation: A review and a research agenda. *The Journal of Strategic Information Systems*, Volume 28, Issue 2, Pages 118-144, https://doi.org/10.1016/j.jsis.2019.01.003.

Schmarzo, B. (2019, November 30). "*What is Digital Transformation?*" https://www.cio.com/article/3199030/what-is-digital-transformation.html

MacDonald, M. P. (2012, November 19). "*Digital Strategy Does Not Equal IT Strategy.*" https://hbr.org/2012/11/digital-strategy-does-not-equa

Wren, H. (2021, September 16). "*What is digital transformation? Definition, examples & importance.*" https://www.zendesk.com/blog/digital-transformation/

Berman, Saul J. (2012, March 1). Digital transformation: opportunities to create new business models. *Strategy & leadership : a publication of Strategic Leadership Forum*, Vol. 40, Issue 2, pages 16 - 24, https://www.emerald.com/insight/content/doi/10.1108/10878571211209314/full/html

Hood, C. (2016, February 12). "*The Concept of Customer Transformation.*" https://chrishood.com/the-concept-of-customer-transformation/

Gilbert, S. J. (2010, February 16). "*The Outside-In Approach to Customer Service.*" https://hbswk.hbs.edu/item/the-outside-in-approach-to-customer-service

Memon, O. (2023, January 31). "*Confusing Quantification: How Many Parts Does An Airliner Have?*" https://simpleflying.com/airliners-how-many-parts/

Ash, A. (2020, August 12). "*The rise and fall of Blockbuster and how it's surviving with just one store left.*" https://www.businessinsider.com/the-rise-and-fall-of-blockbuster-video-streaming-2020-1

Ross, L. (2021, July 8). "*7 Companies that Failed to Adapt to Disruption and Paid the Ultimate Price.*" https://www.thomasnet.com/insights/7-companies-that-failed-to-adapt-to-disruption-and-paid-the-ultimate-price/

Zetlin, M. (2019, September 20). "*Blockbuster Could Have Bought Netflix for $50 Million, but the CEO Thought It Was a Joke.*" https://www.inc.com/minda-zetlin/netflix-blockbuster-meeting-marc-randolph-reed-hastings-john-antioco.html

Ariella, S. (2022, November 14). "*37 Incredible Digital Transformation Statistics.*" https://www.zippia.com/advice/digital-transformation-statistics/

Chapter Two

Sutton, R. I., Hoyt, D. (2016, January 6). "*Better Service, Faster: A Design Thinking Case Study.*" https://hbr.org/2016/01/better-service-faster-a-design-thinking-case-study

Vollmer, C. A. H. (2014, October 28). "*The Art of the Possible.*" https://www.forbes.com/sites/strategyand/2014/10/28/the-art-of-the-possible/

Enterprise. (2019, December 25). "*The art of prediction: Can we tell what the future holds?*" https://enterprise.press/stories/2019/12/25/the-art-of-prediction-can-we-tell-what-the-future-holds-9131

Levy, A. (2023, February 1). "*Meta lost $13.7 billion on Reality Labs in 2022 as Zuckerberg's metaverse bet gets pricier.*" https://www.cnbc.com/2023/02/01/meta-lost-13point7-billion-on-reality-labs-in-2022-after-metaverse-pivot.html

Olson, P. (2023, February 1). "*Zuckerberg's New Focus Pulls Meta Back From The Brink.*" https://www.bloomberg.com/opinion/articles/2023-02-02/mark-zuckerberg-and-facebook-s-parent-pull-back-from-the-metaverse

Hays, K. (2023, March 14). "*Mark Zuckerberg's metaverse ambitions are shrinking.*" https://www.businessinsider.com/meta-layoffs-ceo-mark-zuckerberg-moves-away-from-metaverse-ai-2023-3

Teller, A. (2019, February 11). *"Google X Head on Moonshots: 10X Is Easier Than 10 Percent."* https://www.wired.com/2013/02/moonshots-matter-heres-how-to-make-them-happen/

Ignatius, A. (2019, September). *"How Indra Nooyi Turned Design Thinking Into Strategy: An Interview with PepsiCo's CEO."* pages 80 - 85, Harvard Business Review. https://hbr.org/2015/09/how-indra-nooyi-turned-design-thinking-into-strategy

Woolery, E. (2019). *"Design Thinking Handbook."* https://www.designbetter.co/design-thinking/empathize

Google, (2020). *"Build your creative capacity."* https://rework.withgoogle.com/guides/design-thinking/steps/build-your-creative-capacity/

Voltage Control, (2022, May 21). *"8 Great Design Thinking Examples."* https://voltagecontrol.com/blog/8-great-design-thinking-examples/

von Schmieden, K. (2019). *"Feeling in Control: Bank of America Helps Customers to Keep the Change."* https://thisisdesignthinking.net/2018/09/feeling-in-control-bank-of-america-helps-customers-to-keep-the-change/

Interaction Design Foundation, (2017, April 13). *"What is Design Thinking?"* https://www.interaction-design.org/literature/topics/design-thinking

Gibbons, S. (2016, July 31). *"Design Thinking 101."* https://www.nngroup.com/articles/design-thinking/

Pferdt, F. G. (2019, October). *"Design Thinking in 3 steps: How to build a culture of innovation."* https://www.thinkwithgoogle.com/future-of-marketing/creativity/design-thinking-principles/

Lafargue, V. (2016, June 20). *"How To Brainstorm Like A Googler."* https://www.fastcompany.com/3061059/how-to-brainstorm-like-a-googler

Ponomarev, A. (2019, November 15). *"What Is Pretotyping and How Is It Different from Prototyping and building MVP?"* https://medium.com/rocket-startup/what-is-pretotyping-and-how-is-it-different-from-prototyping-and-building-mvp-b44f21611aa

Naiman, L. (2019, June 10). *"Design Thinking as a Strategy for Innovation."* https://www.creativityatwork.com/design-thinking-strategy-for-innovation/

Glinska, M. (2015, January). "*Innovation and Growth: Understanding the Power of Design Thinking.*" https://issuu.com/batteninstitute/docs/designthinking-121814-issuu

Chapter Three

Burkeman, O. (2001, Dec 29). "*The Virtual Visionary.*" https://www.theguardian.com/technology/2001/dec/29/games.academicexperts

Dover Fueling Solutions, (2020, July 16). "*Dover Fueling Solutions Launches Ground-breaking DFS Anthem UX*™." https://www.doverfuelingsolutions.com/post/dover-fueling-solutions-launches-groundbreaking-dfs-anthem-ux

Lopez, M. (2021, May 11). "*Taking Touchless Interactions To The Next Level With Millimeter Wave.*" https://www.forbes.com/sites/maribellopez/2021/05/11/taking-touchless-interactions-to-the-next-level-with-millimeter-wave/

Iqbal, M. Z., Campbell, A. G. (2021, October 27). "*From luxury to necessity: Progress of touchless interaction technology.*" https://www.ncbi.nlm.nih.gov/pmc/articles/PMC9595506/

Simmons, A. (2023, January 1). "*Internet of Things (IoT) Examples by Industry in 2023.*" https://dgtlinfra.com/internet-of-things-iot-examples/

TelecomNews, (2020, November 20). "*At 12 billion, IoT connections to surpass non-IoT devices in 2020.*" https://telecom.economictimes.indiatimes.com/news/at-12-billion-iot-connections-to-surpass-non-iot-devices-in-2020/79318722

LoveMyEV.com, (2023, June 20). "*On test: The eight best EV route planner apps.*" https://lovemyev.com/explore/charging/on-test-the-six-best-route-planning-apps-for-ev-drivers

Dall'asen, N. (2020, July 23). "*Ulta Beauty's App Can Give You a Skin-Care Routine Based on Facial Recognition.*" https://www.allure.com/story/ulta-beauty-app-skin-analysis-review

Hood, C. (2019, September 7). "*The New API - Application People Interfaces.*" https://chrishood.com/the-new-api-application-people-interfaces/

Porter, E. (2023, January 20). "*Technology Needs More Humanity.*" https://www.washingtonpost.com/business/technology-needs-more-humanity/2023/01/20/34462238-98bc-11ed-a173-61e055ec24ef_story.html

Murayama, R., Swift, R. (2021, December 23). *"Tasty TV: Japanese professor creates flavourful screen."* https://www.reuters.com/technology/lick-it-up-japan-professor-creates-tele-taste-tv-screen-2021-12-23/

Chapter Four

Turing, A. M. (1950, October). *"Computing Machinery and Intelligence."* https://www.turing.org.uk/scrapbook/test.html

Fraser, G. (2014, June 13). *"A computer has passed the Turing test for humanity – should we be worried?"*, https://www.theguardian.com/commentisfree/belief/2014/jun/13/computer-turing-test-humanity

Thomas, M. (2023, March 3). *"The Future of AI: How Artificial Intelligence Will Change the World."* https://builtin.com/artificial-intelligence/artificial-intelligence-future

AP, (2019, April 23). *"Coming to store shelves: cameras that guess your age and sex."* https://apnews.com/article/bc0080f3cf4f4eae9f886ec7dfcd5235

Cuni-Mertz, L.,Jung, H. (2021, November 1). *"What Are the Risks for Retailers When Adopting Artificial Intelligence?"* https://walton.uark.edu/insights/posts/what-are-the-risks-for-retailers-when-adopting-artificial-intelligence.php

Guha, A. et al. (2021, March). How artificial intelligence will affect the future of retailing. *Journal of Retailing*, Volume 97, Issue 1, Pages 28 - 41, https://doi.org/10.1016/j.jretai.2021.01.005

Ventura, R. (2023, January 24). *"Customer Relationships Reshaped by Artificial Intelligence."* https://consumergateway.org/2023/01/24/customer-relationships-reshaped-by-artificial-intelligence/

Libai, B. et al. (2020). Brave New World? On AI and the Management of Customer Relationships. *Journal of Interactive Marketing*, Volume 51, Issue C, Pages 44 - 56, http://www.sciencedirect.com/science/article/pii/S1094996820300839

Ocampo, F. (2017, December 19). *"Frictionless is 2018's Biggest Customer Experience Buzzword."* https://www.openaccessbpo.com/blog/frictionless-2018s-biggest-customer-experience-buzzword/

Smith, A. (2018, March 12). *"Three Ways AI Can Help Build Customer Relationships."* https://www.forbes.com/sites/anthonysmith/2018/03/12/three-ways-ai-can-help-build-customer-relationships

Castillo, D. et al. (2020, June). The dark side of AI-powered service interactions: exploring the process of co-destruction from the customer perspective. *The Service Industries Journal*, Volume 41, Pages 1 - 26, 10.1080/02642069.2020.1787993. https://www.researchgate.net/publication/342574519_The_dark_side_of_AI-powered_service_interactions_exploring_the_process_of_co-destruction_from_the_customer_perspective

Ciechanowski, L. et al. (2019, March). In the shades of the uncanny valley: An experimental study of human–chatbot interaction. *Future Generation Computer Systems*, Volume 92, Pages 539 - 548, https://doi.org/10.1016/j.future.2018.01.055

Demoulin, N., Willems, K. (2019, November). Servicescape irritants and customer satisfaction: The moderating role of shopping motives and involvement. *Journal of Business Research*, Volume 104, Pages 295 - 306, https://doi.org/10.1016/j.jbusres.2019.07.004

Puthiyamadam, T., Reyes, J. (2018). *"Experience is everything: Here's how to get it right."* https://www.pwc.com/us/en/advisory-services/publications/consumer-intelligence-series/pwc-consumer-intelligence-series-customer-experience.pdf

Sklar, E. (2019, March 4). *"Misuse of AI can destroy customer loyalty: here's how to get it right."* https://www.comparethecloud.net/articles/misuse-of-ai-can-destroy-customer-loyalty-heres-how-to-get-it-right/

Gosline, R. R. (2022, June 9). *"Why AI Customer Journeys Need More Friction."* https://hbr.org/2022/06/why-ai-customer-journeys-need-more-friction

Chapter Five

Hansen, D. (2013, December 19). *"Myth Busted: Steve Jobs Did Listen To Customers."* https://www.forbes.com/sites/drewhansen/2013/12/19/myth-busted-steve-jobs-did-listen-to-customers/

Spangler, T. (2023, February 8). *"Disney+ Drops 2.4 Million Subscribers in First Loss, Bob Iger Heralds 'Significant Transformation' Underway."* https://variety-com.cdn.ampproject.org/c/s/variety.com/2023/biz/news/disney-q1-2023-earnings-bob-iger-disney-plus-loses-subscribers-1235517007/amp/

Gilbert, S. J. (2010, February 16). *"The Outside-In Approach to Customer Service."* https://hbswk.hbs.edu/item/the-outside-in-approach-to-customer-service

Chevalier, S. (2022, October 14). *"D2C e-commerce sales in the U.S. 2019-2024."* https://www.statista.com/statistics/1109833/usa-d2c-ecommerce-sales/

von Ahn, T. (2023, April 24). *"Inside-Out Strategy vs. Outside-In Strategy: Which Marketing Approach Is Best?"* https://viralsolutions.net/inside-out-strategy-vs-outside-in-strategy/

Alonso, T. (2023, February 16). *"How Uber Disrupted An Industry With An Explosive Approach."* https://www.cascade.app/studies/uber-strategy-study

Fowler, S. (2017, February 19). *"Reflecting On One Very, Very Strange Year At Uber."* https://www.susanjfowler.com/blog/2017/2/19/reflecting-on-one-very-strange-year-at-uber

Wong, J. C. (2017, June 21). *"Uber CEO Travis Kalanick resigns following months of chaos."* https://www.theguardian.com/technology/2017/jun/20/uber-ceo-travis-kalanick-resigns

Northsail Software. (2021, October 23). *"Customer Data: Your Most Valuable Resource."* https://northsail.io/horizon/articles/5-methods-for-collecting-customer-data

Chapter Six

Poetry Foundation, (2023). *Robert Frost.* https://www.poetryfoundation.org/poets/robert-frost

Weeks, C. (2020, July 30). *"The Power of a Connected Customer Journey."* https://www.forbes.com/sites/forbestechcouncil/2020/07/30/the-power-of-a-connected-customer-journey/

Chapter Seven

Adobe, (2019, April 10). *"Channel-Less Marketing: Why your Brand Needs Channel-Less Communication."* https://business.adobe.com/blog/basics/channel-less-customer-experience-5-characteristics

Nalawade, S. (2021, December 16). *"What is Channel-less Customer Experience and How to Ace It."* https://www.spiceworks.com/marketing/customer-experience/articles/what-is-channelless-customer-experience/

Lucas, C. (2021, January 26). *"The ultimate channel strategy is channel-less."* https://www.odigo.com/blog-and-resources/blog/the-ultimate-channel-strategy-is-channel-less/

Hill, G. (2021, July 21). "*Why you need to map the customer's ecosystem.*" https://www.mycustomer.com/customer-experience/engagement/why-you-need-to-map-the-customers-ecosystem

Taylor, M. (2021, August 3). "*The Customer Experience Ecosystem, Now In 3D.*" https://www.forbes.com/sites/cognizant/2021/08/03/the-customer-experience-ecosystem-now-in-3d/

Adastra. (2022, October 6). "*Customer 720: Driving Actionable Insight From Your Master Data Management System.*" https://adastracorp.com/insights/customer-720-driving-actionable-insight-from-your-master-data-management-system/

Brush, K. (2023, June). "*Digital Ecosystem.*" https://www.techtarget.com/searchcio/definition/digital-ecosystem

Newman, L. (2022, August 23). "*6 Features that Make the Apple Ecosystem Amazing.*" https://www.makeuseof.com/features-that-make-apple-ecosystem-amazing/

Deloitte. (2021, April). "*API-enabled digital ecosystems.*" (PDF) https://www2.deloitte.com/content/dam/Deloitte/in/Documents/Consulting/in-consulting-api-thought-leadership-noexp.pdf

Tao, A. et al. (2022, July 27). "*Open-Source Innovation.*" https://encyclopedia.pub/entry/25442

Haff, G. (2022, November 16). "*How open source powers innovation.*" https://opensource.com/article/22/11/open-source-innovation

Sanaullah, A. (2022, May). "*RISC-V for FPGAs: benefits and opportunities.*" https://research.redhat.com/blog/article/risc-v-for-fpgas-benefits-and-opportunities/

Jurek, M. (2019, June 15). "*Why are Nike Shoes so Darn Popular?*" https://medium.com/@michelejurek16/why-are-nike-shoes-so-darn-popular-fa3b5426730e

Zimmerman, E. (2015, July 1). "*Makers of Custom Sneakers Cash In on Desire for One-of-a-Kind Goods.*" https://www.nytimes.com/2015/07/02/business/smallbusiness/makers-of-custom-sneakers-cash-in-on-desire-for-one-of-a-kind-goods.html

Potepalova, M. (2016, June 17). "*Nike Learns to Mass Customize Shoes While Near-Shoring.*" https://scm.ncsu.edu/scm-articles/article/nike-learns-to-mass-customize-shoes-while-near-shoring

Chapter Eight

Bussgang, J., Bacon, J. (2020, January 21). "*When Community Becomes Your Competitive Advantage.*" https://hbr.org/2020/01/when-community-becomes-your-competitive-advantage

Wilson, G. (2022, August 7). "Go-to-Community Strategy (GTC)." https://orbit.love/article/go-to-community-strategy-gtc

Villegas, M. L. (2022, January 22). "*Digital communities and their impact on the digital ecosystem.*" https://axiacore.com/blog/digital-communities-and-their-impact-on-the-digital-ecosystem-812/

Luzio, C. (2019, December 9). "*Advice for scaling your business through community.*" https://mileiq.com/blog-en-us/advice-scaling-your-business-through-community

Rajan, V. (2017, October 30). "*How to Scale your Customers and Clients into a Community.*" https://www.forbes.com/sites/forbescoachescouncil/2017/10/30/how-to-scale-your-customers-and-clients-into-a-community/

Mingo, C. (2018, November 9). "*The Importance & Value of Business Networking.*" https://powerdigitalmarketing.com/blog/the-importance-value-of-business-networking/

Tanriseven, T. (2022, August 9). "*How to Scale a Flourishing Online Community: 11 Strategies.*" https://businesscollective.com/how-to-scale-a-flourishing-online-community-11-strategies/index.html

Stream Scheme, (2022, December 15). "*What is Twitch? A Brief Overview and History.*" https://www.streamscheme.com/twitch-a-brief-overview-and-history/

Emmonds, J. (2019, October 16). "*What are peer to peer bonuses, and do they work?*" https://www.vestd.com/blog/what-are-peer-to-peer-bonuses-and-do-they-work

Pofeldt, E. (2023, June 27). "*Bored at work Because you're Not Allowed to Innovate? Maybe it's Time to Find a Company that Embraces Backstage Creativity.*" https://www.forbes.com/sites/elainepofeldt/2023/06/27/bored-at-work-because-youre-not-allowed-to-innovate-maybe-its-time-to-find-a-company-that-embraces-backstage-creativity/

Chapter Nine

Waters, S. (2022, July 19). "*Why social capital might be the most valuable asset you aren't using.*" https://www.betterup.com/blog/social-capital

Vantage Circle. (2023, April 27). "*The Role of HR in Employee Relationship Management (ERM).*" https://blog.vantagecircle.com/employee-relationship-management/

Wooll, M. (2021, July 1). "*The importance of listening as a leader in the digital era.*" https://www.betterup.com/blog/the-importance-of-listening-as-a-leader-in-the-digital-era

Jurisic, N. et al (2020, August 4). "*Doing vs being: Practical lessons on building an agile culture.*" https://www.mckinsey.com/capabilities/people-and-organizational-performance/our-insights/doing-vs-being-practical-lessons-on-building-an-agile-culture

Axe, J. (2022, August 31). "*Servant Leadership: The Ultimate Key To A Healthy Business.*" https://www.leaders.com/articles/leadership/servant-leadership/

Gentry, W. A. et al (2011, November). "*Empathy in the Workplace.*" (PDF) https://cclinnovation.org/wp-content/uploads/2020/03/empathyintheworkplace.pdf

Levy, S. (2023, January 27). "*Alphabet's Layoffs Aren't Very Googley.*" https://www.wired.com/story/plaintext-alphabets-layoffs-arent-very-googley/

Hood, C. (2021, September 27). "*What is a Culture of Innovation for a Business?*" https://chrishood.com/what-is-a-culture-of-innovation-for-a-business/

The Washington Post. (2021, May 24). "*How better collaboration can boost innovation and success in the new normal.*" https://www.washingtonpost.com/brand-studio/wp/2021/05/24/feature/how-better-collaboration-can-boost-innovation-and-success-in-the-new-normal/

McCleary, T. (2015, January 15). "*Good to Great: Validate Your Employees.*" https://fowmedia.com/good-great-validate-employees/

Chapter Ten

Vereckey, B. (2021, December 21). "*4 Strategies for Digital Growth from Spotify's CFO.*" https://mitsloan.mit.edu/ideas-made-to-matter/4-strategies-digital-growth-spotifys-cfo

Robinson, K. (2021). "*15 Years of Spotify: How the Streaming Giant Has Changed and Reinvented the Music Industry.*" https://variety.com/2021/music/news/spotify-turns-15-how-the-streaming-giant-has-changed-and-reinvented-the-music-industry-1234948299/

Šmite, D. et al (2023). Decentralized decision-making and scaled autonomy at Spotify. *Journal of Systems and Software*, Volume 200, https://doi.org/10.1016/j.jss.2023.111649

McKinsey. (2017, January 10). "*ING's Agile Transformation.*" https://www.mckinsey.com/industries/financial-services/our-insights/ings-agile-transformation

Agile Business Consortium, (2022, November 28). "*Case Study: ING's Digital Platform Tribe Goes Agile.*" https://www.agilebusiness.org/resource-report/case-study-ings-digital-platform-tribe-goes-agile.html

Brand Finance, (2015, February 17). "*Lego Overtakes Ferrari as the World's Most Powerful Brand.*" https://brandfinance.com/press-releases/lego-overtakes-ferrari-as-the-worlds-most-powerful-brand

Merrill, P. (2022, June 15). "*The cautionary and inspirational story of how LEGO rebuilt itself.*" https://www.theceomagazine.com/business/company-profile/rebuilding-lego/

Toren, M. (2013, June 21). "*Lego's Secrets for Brand Longevity.*" https://www.entrepreneur.com/business-news/legos-secrets-for-brand-longevity/229013

Linders, B. (2017, September 28). "*Agile at LEGO.*" https://www.infoq.com/news/2017/09/agile-lego/

Slater, D. (2022, November 2). "*Powering Innovation and Speed with Amazon's Two-Pizza Teams.*" https://aws.amazon.com/executive-insights/content/amazon-two-pizza-team/

Haden, J. (2021, February 10). "*When Jeff Bezos's 2-Pizza Teams Fell Short, He Turned to the Brilliant Model Amazon Uses Today.*" https://www.inc.com/jeff-haden/when-jeff-bezoss-two-pizza-teams-fell-short-he-turned-to-brilliant-model-amazon-uses-today.html

Clear Bridge Mobile, (2020, August 15). "*Agile Transformation Challenges: 6 Missteps That Slow Down Change.*" https://clearbridgemobile.com/agile-transformation-challenges/

Walt Disney Imagineering, (2023). "*Outreach and Inspiration.*" https://sites.disney.com/waltdisneyimagineering/our-impact/

Chapter Eleven

Tran, D. (2023, June 6). *"Where Is Uprising Bread From Shark Tank Today?"* https://www.thedailymeal.com/1198381/where-is-uprising-bread-from-shark-tank-today/

Zhao, M. S. (2021, December 5). *"Kristin and William Schumacher of Uprising Food: 5 Things You Need To Create a Successful Food Brand."* https://medium.com/authority-magazine/kristin-and-william-schumacher-of-uprising-food-5-things-you-need-to-create-a-successful-food-5a0b54c353cf

Uprising Food, (2021, October 13). *"Uprising Food Wins on Taste with Their Superfood Bakery Product Line on Season Premiere of Shark Tank."* https://www.prnewswire.com/news-releases/uprising-food-wins-on-taste-with-their-superfood-bakery product-line-on-season-premiere-of-shark-tank-301399256.html

Carstea, M. (2022, August 9). *"Purpose driven brands lead on business performance."* https://www.gfk.com/blog/purpose-driven-brands-lead-on-business-performance-gfk-blog

Lugtu, R. (2023, April 27). *"Purpose-Driven Technology."* https://www.institutefordigitaltransformation.org/purpose-driven-technology/

RevBoss, (2022, September 7). *"How a Purpose-Driven Culture Gets you More Loyal Customers."* https://revboss.com/blog/how-a-purpose-driven-culture-gets-you-more-loyal-customers

Corporate Finance Institute, (2020, April 29). *"Kantian Ethics."* https://corporatefinanceinstitute.com/resources/esg/kantian-ethics/

C-Span, (2023, March 23). *"TikTok CEO Testifies at House Energy and Commerce Committee Hearing."* https://www.c-span.org/video/?526609-1/tiktok-ceo-testifies-house-energy-commerce-committee-hearing

Watters, A. (2023, February 3). *"5 Ethical Issues in Technology to Watch for in 2023."* https://connect.comptia.org/blog/ethical-issues-in-technology

Scherson, A. (2022, December 5). *"The Importance of Being a Transparent Leader - and How to Be One."* https://www.uschamber.com/co/grow/thrive/how-to-lead-with-transparency

Hood, C. (2019, July 24). *"A Thesis for Digital Acceleration."* https://chrishood.com/a-thesis-for-digital-acceleration/

Serheichuk, N. (2021, September 2). *"Digital Acceleration: What, why, and how."* https://www.n-ix.com/digital-acceleration/

Cone, C. (2019, July 28). *"What does a Purpose-Driven Company Look Like?"* https://www.salesforce.org/blog/what-does-a-purpose-driven-company-look-like/

Firsau, M. (2023, April 14). *"How To Become A Purpose-Driven Business Consumers Trust."* https://www.forbes.com/sites/forbesbusinesscouncil/2023/04/14/how-to-become-a-purpose-driven-business-consumers-trust/

Hannum, L. (2020, April 11). *"Engaging Employees in Purpose, Mission and Values Through Strategic Communication."* https://beehivepr.biz/engaging-employees-in-purpose-mission-and-values-through-strategic-communication/

Harter, J. (2018, August 26). *"Employee Engagement on the Rise in the U.S."* https://news.gallup.com/poll/241649/employee-engagement-rise.aspx

Goel, S. (2022, November 12). *"How does Duolingo work & make money: Business Model & Strategy."* https://thestrategystory.com/2022/11/12/how-does-duolingo-work-make-money-business-model-strategy/

Mascarenhas, N. (2021, May 3). *"How Duolingo became fluent in monetization."* https://techcrunch.com/2021/05/03/duolingo-ec1-monetization/

Pressman, M. (2021, July 19). *"Elon Musk Is 'Pioneer' Of Purpose-Driven Brands, Says New Study."* https://evannex.com/blogs/news/elon-musk-considered-pioneer-of-purpose-driven-brands-according-to-new-study

Taylor, E. et al (2020, July 22). *"How Tesla defined a new era for the global auto industry."* https://www.reuters.com/article/us-autos-tesla-newera-insight/how-tesla-defined-a-new-era-for-the-global-auto-industry-idUSKCN24N0GB

Johnson, T. (2022, September 19). *"Zipline Drones Deliver Medical Supplies Across Africa."* https://borgenproject.org/zipline-drones/

Kai, K. (2023, April 28). *"Drone Delivery Startup Zipline Boosts Valuation to $4.2 Billion."* https://www.forbes.com/sites/alexkonrad/2023/04/28/drone-delivery-startup-zipline-boosts-valuation-to-4-billion/

O'Brien, D. et al. (2019, October 15). "*Purpose is Everything.*" https://www2.deloitte.com/us/en/insights/topics/marketing-and-sales-operations/global-marketing-trends/2020/purpose-driven-companies.html

Chapter Twelve

Stobierski, T. (2019, August 26). "*The Advantages of Data-Driven Decision-Making.*" https://online.hbs.edu/blog/post/data-driven-decision-making

McCloskey, H. (2022, January 16). "*The Human Element of Data Driven Decision Making.*" https://uservoice.com/blog/data-driven-decision-making

Garvin, D. A. (2013, December). "*How Google Sold Its Engineers on Management.*" https://hbr.org/2013/12/how-google-sold-its-engineers-on-management

MacKenzi, I. et al. (2013, October 1). "*How retailers can keep up with consumers.*" https://www.mckinsey.com/industries/retail/our-insights/how-retailers-can-keep-up-with-consumers

Hansen, H. L. (2019, April 29). "*In God we trust. All others must bring data.*" https://www.ibm.com/blogs/nordic-msp/in-god-we-trust-all-others-must-bring-data/

Brown, C. (2022, March 15). "*Executing a Data Strategy with OKRs.*" https://towardsdatascience.com/executing-a-data-strategy-with-okrs-acbdbbf126a7

Voice123, (2022, May 16). "*OKRs and KPIs: the methodology for appropriate decision making.*" https://voice123.com/about/our-handbook/okr-and-kpi/

Van Veenendal, M (2022, September 5). "*Digital Acceleration: A Present and Future Necessity.*" https://www.codepwr.com/digital-acceleration-a-present-and-future-necessity/

Khanna, A. (2022, March 4). "*Five Ways Decision Intelligence Helps Accelerate Data-Driven Decision-Making.*" https://www.dataversity.net/five-ways-decision-intelligence-helps-accelerate-data-driven-decision-making/

Rimol, M. (2021, September 13). "*Gartner Survey Reveals Talent Shortages as Biggest Barrier to Emerging Technologies Adoption.*" https://www.gartner.com/en/newsroom/press-releases/2021-09-13-gartner-survey-reveals-talent-shortages-as-biggest-barrier-to-emerging-technologies-adoption

March, L. (2022, November 24). *"Data-Driven Decision-Making: Unlocking Sustainable Success."* https://www.similarweb.com/blog/research/market-research/data-driven-decision-making/

Bean, R. (2022, January 3). *"NewVantage Partners Releases 2022 Data And AI Executive Survey."* https://www.businesswire.com/news/home/20220103005036/en/NewVantage-Partners-Releases-2022-Data-And-AI-Executive-Survey

ISHIR, (2023, May 9). *"Data-driven decision-making: How to use data analytics to drive business decisions."* https://securityboulevard.com/2023/05/data-driven-decision-making-how-to-use-data-analytics-to-drive-business-decisions/

Chapter Thirteen

Geraghty, J. (2023, April 13). *"Bud Light Doesn't Seem to Understand Its Own Consumers."* https://www.nationalreview.com/the-morning-jolt/bud-light-doesnt-seem-to-understand-its-own-consumers/

Graziano, D. (2023, May 28). *"Beer and Backlash: Lessons From The Bud Light Mess."* https://chiefexecutive.net/beer-and-backlash-lessons-from-the-bud-light-mess/

Coggins, M (2023, May 19). *"Former Anheuser-Busch exec says Bud Light backlash not 'going away'."* https://www.foxbusiness.com/media/former-anheuser-busch-exec-bud-light-backlash-not-going-away

Zilber, A. (2023, April 13). *"'No one at senior level' of Bud Light knew of Dylan Mulvaney ad campaign."* https://nypost.com/2023/04/13/no-one-at-senior-level-of-bud-light-knew-of-dylan-mulvaney-ad-campaign/

Peters, B. (2023, April 14). *"Anheuser-Busch CEO on the Bud Light backlash: 'We never intended to be part of a discussion that divides people'."* https://www.marketwatch.com/story/anheuser-busch-ceo-on-the-bud-light-backlash-we-never-intended-to-be-part-of-a-discussion-that-divides-people-9a68a6fc

Whitworth, B. (2023, June 15). *"Anheuser-Busch Announces Support For Frontline Employees And Wholesaler Partners."* https://www.anheuser-busch.com/newsroom/support-for-frontline-employees-and-wholesaler-partners

Koberg, K. (2023, June 11). "*Bud Light demand has 'plummeted completely' since Dylan Mulvaney controversy: Bartending company founder.*" https://www.foxnews.com/media/bud-light-demand-plummeted-completely-dylan-mulvaney-controversy-bartending-company-founder

James, E. (2023, June 7). "*Modelo Especial is now the number one selling beer - as Bud Light sales fell to $297 million.*" https://www.msn.com/en-us/money/companies/modelo-especial-is-now-the-number-one-selling-beer-as-bud-light-sales-fell-to-dollar297-million/

VanBoskirk, S. (2021, August 3). "*How And Why To Bother With Customer Obsession.*" https://www.forrester.com/blogs/how-and-why-to-bother-with-customer-obsession/

Amire, R. (2022, May 10). "*Purpose at Work Predicts if Employees Will Stay or Quit Their Jobs.*" https://www.greatplacetowork.com/resources/blog/purpose-at-work-predicts-if-employees-will-stay-or-quit-their-jobs

Western Governors University, (2021, January 15). "*What is adaptive leadership?*" https://www.wgu.edu/blog/what-adaptive-leadership2101.html

All Activity, (2022, April 7). "*Purpose-Driven Leadership: A Complete Guide.*" https://allactivity.com/blog/purpose-driven-leadership-succeed/

Hannan, M. T. (2007, March 22). "*Organizational Analysis.*" https://www.britannica.com/money/topic/organizational-analysis

Burns, J. M. (1978). *Leadership*. Harper & Row, New York. https://www.worldcat.org/title/Leadership/oclc/3632001

University of Massachusetts, (2021, September 22). "*What is transformational leadership? Understanding the impact of inspirational guidance.*" https://www.umassglobal.edu/news-and-events/blog/what-is-transformational-leadership

Cherry, K. (2023, February 24). "*How Transformational Leadership Can Inspire Others.*" https://www.verywellmind.com/what-is-transformational-leadership-2795313

Runyon, M. (2021, November 18). "*How active listening can make you a better leader.*" https://enterprisersproject.com/article/2021/11/how-active-listening-can-make-you-better-leader

Ghosh, U. (2023, May 28). *How $1,000,000 Cold-Call Helped Mark Cuban Overcome Mavericks' New CEO's 'S*xual Assault' Worries: Culture Transformation and Leadership.* https://thesportsrush.com/nba-news-how-1-million-cold-call-helped-mark-cuban-overcome-mavericks-new-ceos-sxual-assault-worries-culture-transformation-and-leadership/

Chapter Fourteen

Alfred, L. (2022, January 3). "*7 Key Principles of Value-Based Selling.*" https://blog.hubspot.com/sales/value-based-selling

Maza, C. (2023, May 23). "*What is customer value? Definition, formula & importance.*" https://www.zendesk.com/blog/customer-value/

Brown, L. (2022, October 3). "*What is Business Value and How it is Measured?*" https://www.invensislearning.com/blog/what-is-business-value/

Boyles, M. (2022, March 8). "*Innovation in Business: What it is & Why it's so Important.*" https://online.hbs.edu/blog/post/importance-of-innovation-in-business

Richards, K. (2015, March 23). "*Corporate Culture - Values Alignment is the Foundation.*" https://www.huffpost.com/entry/corporate-culture-value_b_6927466

Ronan, A. (2021, March 15). "*5 Ways to Align Your Business and Your Values.*" https://www.gcu.edu/blog/business-management/5-ways-align-your-business-and-your-values

Eberlin, J. (2013, July 3). "*Implementing a Vibrant Online Community.*" https://www.gainsight.com/customer-success-best-practices/implementing-a-vibrant-online-community/

INDEX

10x Thinking, 38, 40, 49
Adam Walendziewski, 189
adaptability, 135, 174, 175, 186
Adaptability, 170
adaptive leadership, 242
advertising, 70, 95, 206, 238
Agile, 49
AI, 67, 68, 69, 70, 71, 72, 73, 74, 75, 76, 77, 78, 79, 80, 81, 82, 83, 84, 110, 117, 130, 133, 134, 207, 209, 222, 225, 227, 228
Ajay Khanna, 226
Alan Turing
Turing, 67
Albert Einstein, 34
Alberto Savoia
Savoia, 47
Alexa, 56
algorithm, 70, 158
Alicia Paterson, 46
Paterson, 46, 47
Alissa Heinerscheid, 236
All Activity, 243
Amazon, 74, 156, 192, 193, 221
Amit Zavery, 21, 33
Anaheim, 11
analytics, 78, 136, 175, 176, 209, 212, 220, 221, 222, 226, 227
Anheuser-Busch, 235, 236, 238
Anson Frericks, 236
Anthony Smith, 74, 80
API
Application Programming Interface, 135
Apple, 125, 134, 146
AromaRama, 62
Arthur C. Clarke, 125
artificial intelligence, 79, 134, 136, 212
Atlassian, 180
Augmented Reality, 117
Automation, 227
B2B, 93, 94, 95
B2C, 93, 94
backstage creativity, 158, 160
Barbara Corcoran, 201
Barnastics, 239
Bart Schlatmann, 190

basketball, 251
Bernard Bass, 245
bias
confirmation bias
organizational bias, 176, 219
big-picture thinking, 37, 38
Bill Schmarzo, 24
Blockbuster, 31, 32
Blue Ocean, 49
Bob Iger, 91
Boeing, 30
brand reputation, 98, 258, 272
Brendan Whitworth, 237
Bud Light, 235, 236, 237, 238, 239, 251
Budweiser, 236
business leaders, 261, 273
Business Value, 3, 4, 257, 260
California, 3, 4, 11, 105, 106, 107
Cameron Weeks, 113
Carol Cone, 210
Carrefour, 135
Case Study, 156, 237
Catarina Tucker, 239
Cate Luzio, 147
CEO, 91, 100, 147, 151, 191, 206, 219, 221, 237, 249, 250, 251, 261
Change, 186
ChatGPT, 159
Chuck Gulledge, 41
Cinemapolis, 11, 13
CIO, 78, 79, 80, 94, 95
Claire Lucas, 127
Clearbridge Mobile, 193
Cloud, 135, 136, 254
Collaboration, 133, 171, 179, 180, 209
community, 96, 117, 136, 142, 143, 145, 146, 147, 148, 149, 150, 151, 152, 153, 154, 155, 156, 157, 167, 172, 175, 177, 186, 189, 210, 211, 216, 221, 258, 259, 262, 268, 272
Continuous improvement, 228
COVID-19, 208
CTO, 67, 255, 256
CTVA, 263, 264, 272, 273

INDEX 297

Cue Health, 63, 64
Culture, 192
Customer 720, 129, 130
Customer-centric. See also customer-centricity
customer-centricity, 182, 220, 237, 239, 257
Customer Ecosystems, 128
customer experiences, 78, 84, 114, 117, 134, 167, 168, 173, 176, 179, 181, 182, 207, 241, 242, 244, 246, 248, 259, 261, 272
customer-first, 95, 188
customer journey, 83, 95, 113, 114, 115, 116, 119, 121, 133, 227, 258, 261, 267
customer lifetime value, 120, 240, 260
customer-obsessed, 241, 243, 247, 248, 249, 251, 263
Customer Obsession, 239, 240, 241
customer satisfaction score, 260
Customer Transformation, 68, 70, 74, 78, 84, 92, 100, 102, 107, 115, 119, 120, 126, 131, 143, 147, 167, 169, 175, 179, 183, 187, 210, 216, 239, 243, 245, 247, 248, 253, 259, 260, 261, 264, 272
Customer Transformation and Value Alignment, 263
Customer Value, 253
customer value propositions, 203
CVS, 210, 211
Cynt Marshall, 251
Dallas Mavericks, 251
Daniel Slater, 192
Danske Bank, 134
data, 4, 7, 79, 80, 175, 209, 217, 220, 221, 223, 224, 226, 227, 228
data-driven, 171, 175, 176, 209, 220, 221, 223, 225, 228, 229, 250
Data-Driven Decision Making, 220, 221
David Vandegrift, 67
DDDM, 220, 221, 223, 224, 225, 226, 227, 228, 229
Denise Graziano, 236
Design Thinking, 42, 45, 49
Devils Tower, 105, 107
digital acceleration, 174, 208, 209, 210, 224, 225, 226, 259

Digital Acceleration, 224
digital ecosystem, 133, 134, 135, 139, 151, 153, 155
digital transformation, 26, 27
Disney, 91, 92, 93, 146, 184, 185, 194
Disneyland, 185, 186
Dominic Price, 180
Duolingo, 209, 212, 213, 214
Dylan Mulvaney, 235
ego, 249
Eli Woolery, 44
Elon Musk, 214
Emma Grede, 202
Empathy, 169, 171, 178, 182
employee, 75, 127, 167, 169, 170, 172, 173, 174, 175, 177, 178, 179, 180, 181, 182, 183, 186, 189, 194, 211, 212, 216, 220, 225, 240, 245, 246, 248, 259, 262, 272
Employee alignment, 262
Eric Teller, 39
Ethical
ethical experiences
ethics, 206
EV, 57
feedback, 101, 102, 103, 104, 121, 147, 148, 155, 171, 174, 176, 177, 179, 181, 183, 207, 211, 225, 228, 249, 250, 263
football, 237
Forbes, 80, 113, 149, 153
Forrester, 239, 240
Fortune, 12
Frederiek G. Pferdt, 46
gaming industry, 143, 157
Gareth Wilson, 144
Gartner, 75, 226
GGRC, 34, 35, 37
Ghana, 215
GirlsAskGuys, 151
Goldilocks, 264
good friction, 83
Google, 21, 39, 42, 46, 47, 135, 136, 146, 159, 220, 254, 281, 292, 297
Google Cloud, 135, 136, 254
Gordon Haff, 137
Graham Hill, 128, 131
Gwen Stefani, 13
Harley Davidson, 152

298 INDEX

Harvard, 83, 92, 102, 143, 167, 220, 241
Helen Keller, 141
Hewlett Packard, 157
Home Depot, 135, 201
Homei Miyashita, 62
House Energy and Commerce Committee, 206
Howard Schultz, 221
HR, 175
human engagement
human
  human values, 81
IDEO, 42
Immanuel Kant, 206
Inclusion Leadership Award, 252
Indeed, 71, 75, 76, 138, 146, 181, 183, 211, 240
ING Bank, 189, 190
innovate, 138, 148, 180, 189, 192, 210, 258, 266, 268
Innovation
  innovate, 98, 118, 125, 127, 128, 133, 135, 136, 137, 138, 139, 140, 141, 143, 147, 155, 156, 171, 175, 180, 186, 191, 192, 209, 223, 225, 239, 242, 244, 258, 259, 265, 266
insights, 84, 96, 100, 101, 102, 171, 175, 176, 180, 183, 207, 209, 222, 226, 227, 229, 240, 249, 259, 260, 272
Insights, 169, 171, 175, 180, 209
Internet of Things, 56
IoT, 110, 133, 226
James Burns, 245, 246
James Taylor, 158, 160
Jaron Lanier, 55
Jeanne Liedtka, 48
Jeffrey Bussgang, 143
Jono Bacon, 143
Josh Axe, 177
Kendra Cherry, 247
Kevin O'Leary, 202
KPI
  Key Performance Indicators, 224
Lead by example, 249
Leaders, 172, 177, 237, 240, 250, 251, 262
Leadership, 4, 7, 220, 228, 237, 241, 243, 248, 252, 262

Leadership Alignment, 4, 7, 220, 228, 237, 241, 243, 248, 252, 262
Lean Startup, 49
LEGO, 190, 191
Lember Gordon, 203
Lestraundra Alfred, 256
LinkedIn, 31, 130, 146
listen, 177, 181, 238, 247, 249, 268
Lori Greiner, 202
Luis von Ahn, 213
Machine Learning, 227
Major League Baseball
  MLB, 117
Marc Winn, 39
María Lucía Villegas, 144
Mark Cuban, 202, 251
Marketing, 70, 127, 204, 236
Mark Taylor, 129
Marty Linsky, 241
Matt Van Veenendal, 225
Max Firsau, 211
Max Weber, 245
McKinsey, 221
Metrics, 223, 260
Mission, 208
MMORPG, 141
Modelo Especial, 239
Moments, 108
Napoleon Bonaparte, 235
Natasha Mascarenhas, 213
National Review, 235
Natural Language Processing, 227
NBA, 252
NCAA, 235
Net Promoter Score, 103, 260
New York Times, 138
NFL, 236
Nike, 138, 139
NY Post, 237
Obsessed Leadership, 7, 248
Omnichannel, 126
outside-in perspective, 92, 94, 95, 96, 97, 98, 100, 103, 119, 120, 121, 147
Pam Dixon, 69
people, 69, 74, 78, 92, 94, 97, 101, 106, 108, 150, 151, 152, 154, 156, 168, 174, 175, 177, 188, 192, 202, 205, 206, 208, 210, 212, 213, 215, 216, 220, 221, 237, 244, 252, 259

INDEX 299

Personalization, 227
Pete Docter, 97
Pixar, 97
PRAISE, 167, 169, 170, 171, 183, 216
privacy, 69, 70, 71, 72, 82, 83, 111
profit, 185, 204, 205, 206, 210, 213, 214
purpose-driven, 203, 204, 205, 208, 209, 210, 212, 214, 216, 241, 243, 244, 245
purpose-driven leadership, 243, 244
PWC, 78
Rahim Hajee, 130
Ranjay Gulati, 92, 102
Red Rock Café, 105
relationship, 67, 68, 70, 72, 95, 109, 112, 148, 149, 151, 173, 181, 208, 211, 223, 248
Relationships, 7, 169, 170, 171, 172, 179, 182
retention, 131, 148, 174, 212, 220, 228, 258, 259, 260
RevBoss, 205
RISC-V, 137
Robert Frost, 105
ROI, 240, 259, 260
Ronald Heifetz, 241
Ron Ventura
 Ventura, 70
Ryan Calo, 69
SAP, 48
Saul Berman, 25
scale, 128, 155, 180, 188, 213, 214, 227, 248, 264
Schumacher
 Kristen
  William, 201, 202
scrum, 188
Sean Connery, 13
Sean Gerety, 201
Security, 229, 256, 257
Service, 169, 171, 176, 180, 182
Severin Hacker, 213
Shark Tank, 201, 202
Shou Zi Chew, 206
Simon Sinek, 243
Smell-O-Rama, 62
society, 107, 131, 194, 203, 205
Spotify, 187, 188, 189
SQL, 227
stakeholders, 136, 181, 205, 209, 212, 216, 221, 250, 259, 261, 264, 272
Starbucks, 146, 220, 221
Star Trek, 79, 80
Steve Jobs, 91
Steven Spielberg, 105
strategy, 83, 84, 92, 97, 98, 99, 100, 126, 145, 146, 147, 172, 187, 190, 193, 212, 223, 225, 227, 229, 239, 241, 249, 250, 259, 273
Taj Mahal, 62
Tesla, 209, 214
The Honest Company, 211
Theodore Levitt, 253
TikTok, 206, 235
Tim Stobierski
 Stobierski, 220
transactional leadership, 245, 246, 247
transformational leadership, 245, 246, 247, 248
Twitch, 156, 157
Twitter, 146
Uber, 98, 99, 100, 116
Ulta, 57, 58
Uprising Food, 201, 202, 203
Utathya Ghosh, 251
Value Alignment, 7, 261, 263, 264
value-based selling, 256
Variety, 91
Vig Knudstorp, 191
Vikram Rajan, 149
virtual reality, 61
Walgreens, 69, 70, 71
Walt Disney, 184, 185
W. Edwards Deming, 217
Wholesaler, 238
Workshop, 16, 51, 86, 120, 162, 196, 231, 275
Zappos, 159
Zebra Technologies, 64, 66
Zipline, 214, 215

Made in the USA
Monee, IL
03 April 2024